Scrolling Ourselves to Death

Other Gospel Coalition Books

To explore all titles from the Gospel Coalition, including those in the
New City Catechism, TGC Kids, and TGC Hard Questions lines, visit
TGC.org/books.

"Rarely does a collection of chapters from diverse contributors come together to form such a cohesive vision, offering penetrating insights into our current cultural moment. *Scrolling Ourselves to Death* goes beyond merely revisiting Neil Postman's groundbreaking work; it uses Postman's insights as a springboard for deeper reflection and application, all while keeping an eye on the eternal truths of Scripture that remain unchanged in our rapidly advancing technological age."

Trevin Wax, Vice President for Research and Resource Development, North American Mission Board; Visiting Professor, Cedarville University; author, *The Thrill of Orthodoxy, The Multi-Directional Leader*; and *This Is Our Time*

"If you're feeling anxious, irritable, or tired today, one reason is that you've probably spent about five hours on your smartphone—texting, checking the weather, or scrolling social media. In *Scrolling Ourselves to Death*, a slate of authors explains how screens are changing us—and how Christians are uniquely positioned to choose a fuller, better life. By reflecting and building on Neil Postman's insights about television, this book will help you reevaluate and reimagine the choices you're making for yourself, your family, and your community."

Sarah Zylstra, Senior Writer, The Gospel Coalition; editor, *Social Sanity in an Insta World*

"There are books that are enjoyable and books that are important; *Scrolling Ourselves to Death* is both. Although some of the content is sobering—disturbing at times—the contributors never leave the reader hopeless. This is a vitally important book that will help the church clearly communicate the gospel to a world bombarded by distraction."

John Perritt, Director of Resources, Reformed Youth Ministries; author, *Social Media Pressure: Finding Peace Alongside Jesus*

Scrolling Ourselves to Death

Reclaiming Life in a Digital Age

Edited by

Brett McCracken
and Ivan Mesa

WHEATON, ILLINOIS

Scrolling Ourselves to Death: Reclaiming Life in a Digital Age

© 2025 by Brett McCracken and Ivan Mesa Jr.

Published by Crossway
      1300 Crescent Street
      Wheaton, Illinois 60187

Cover design: David Fassett

Cover image: Getty Images and iStock

First printing 2025

Printed in the United States of America

Trade paperback ISBN: 978-1-4335-9944-6
ePub ISBN: 978-1-4335—9946-0
PDF ISBN: 978-1-4335-9945-3

Library of Congress Cataloging-in-Publication Data

Names: McCracken, Brett, 1982– editor. | Mesa, Ivan, editor.
Title: Scrolling ourselves to death : reclaiming life in a digital age / edited by Brett McCracken and Ivan Mesa, Jr.
Description: Wheaton, Illinois : Crossway, 2025. | Series: The gospel coalition | Includes bibliographical references and index.
Identifiers: LCCN 2024032493 (print) | LCCN 2024032494 (ebook) | ISBN 9781433599446 (trade paperback) | ISBN 9781433599453 (pdf) | ISBN 9781433599460 (epub)
Subjects: LCSH: Technological innovations—Religious aspects—Christianity. | Digital media—Religious aspects—Christianity. | Social media—Religious aspects—Christianity. | Christian life.
Classification: LCC BR115.T42 S46 2025 (print) | LCC BR115.T42 (ebook) | DDC 248.4—dc23/eng/20240911
LC record available at https://lccn.loc.gov/2024032493
LC ebook record available at https://lccn.loc.gov/2024032494

Crossway is a publishing ministry of Good News Publishers.

LB   34   33   32   31   30   29   28   27   26   25
15   14   13   12   11   10   9   8   7   6   5   4   3   2

Contents

Introduction

Back to the Future

How a 1985 Book Predicted Our Present

Brett McCracken

HEADS DOWN. Phones out. Fingers scrolling. This is the humanoid posture of our age.

We see it everywhere. Sit in a coffee shop and look around you. All eyes on devices. Wait in line at the post office or grocery store. All eyes on devices. Sit at a red light and look at the drivers in the cars around you. Same story. More disturbing still, look at the drivers on the highway going full speed. Even some of *them* have their eyes darting between the windshields and their smartphones.

We see it in ourselves too. Sit down to read a physical book with your phone nearby. Observe how long you can go without scrolling, texting, or checking some notification. When you're standing in line at a coffee shop and have forty-five seconds to spare, notice how hard it is to resist the urge to pull out your phone to do something—*anything*—to fill that blank space. More disturbing still, monitor how much time elapses between the moment you wake in the morning until the moment you unlock your phone and start scrolling.

For many of us, it's only a matter of seconds.

From the rising of the sun to its going down, we scroll our way through the day. We scroll our way through life. And we are scrolling ourselves to death.

The death march of our scrolling society is not just a metaphor. In many ways, the smartphone is literally killing us (and not just in distracted-driving automobile accidents). Researchers have made compelling correlations between smartphone (especially social media) usage and rising mental unhealth (depression, anxiety, suicidal ideation, loneliness), especially among teens and young adults.[1] Consider the staggering rise in suicide rates among US youth and young adults since the dawn of the smartphone age. Between 2001 and 2007, the suicide rate for kids ages ten to twenty-four was fairly stable, but since 2007 (the year the iPhone debuted), it has skyrocketed, rising 62 percent between 2007 and 2021.[2]

Technology has also helped accelerate a "loneliness epidemic" with demonstrable, wide-ranging negative effects on overall health.[3]

---

1 See especially Jean Twenge, *iGen: Why Today's Super-Connected Kids Are Growing Up Less Rebellious, More Tolerant, Less Happy—and Completely Unprepared for Adulthood* (New York: Atria Books, 2017) and *Generations: The Real Differences between Gen Z, Millennials, Gen X, Boomers, and Silents—and What They Mean for America's Future* (New York: Atria Books, 2023); and Jonathan Haidt, *The Anxious Generation: How the Great Rewiring of Childhood Is Causing an Epidemic of Mental Illness* (New York: Penguin, 2024).

2 Sally C. Curtin and Matthew F. Garnett, "Suicide and Homicide Death Rates among Youth and Young Adults Aged 10–24: United States, 2001–2021," NCHS Data Brief, no. 471, June 2023, https://www.cdc.gov.

3 Tatum Hunter, "Technology's Role in the 'Loneliness Epidemic,'" *Washington Post*, April 11, 2023, https://www.washingtonpost.com/.

The ominous term "deaths of despair" has become part of contemporary vernacular. And after steadily climbing for most of the last century, average life expectancies in the United States have, since 2021, started to decline.

Certainly more than technology is at play in these trends. But not less. When we consider the variables that have *most* changed in society in the last two decades, any answer we come up with will center around digital technology. We didn't know what "social media" was twenty-five years ago. The term *smartphone* was first coined in 1997. The World Wide Web is barely three decades old. Each of these things has utterly reshaped the world in the last quarter century. And things continue to move fast—so fast that we rarely pause long enough to ask questions or ponder unintended side effects. As Antón Barba-Kay put it in *A Web of Our Own Making*, digital technology has so vastly transformed human life over just a few decades that "there is now arguably a greater chasm between someone age twelve and someone age fifty (or forty, or thirty) than there ever was between people separated by a millennium of pharaonic rule in ancient Egypt."[4]

Our critical faculties struggle to keep pace with the scope and speed of the digital revolution. As a result, we're often blind to the ways we're being transformed. If we could jump forward in time a few decades, we could see more clearly. But since we can't do that, our best path to wisdom is often in the other direction: looking back in time, learning from bygone eras and voices. What we can't see now can be illuminated, at least in part, by the insights of generations past.

One book I return to again and again is Neil Postman's *Amusing Ourselves to Death*. The book was prophetic when it released in 1985, and it's even more prophetic now, four decades later.

4 Antón Barba-Kay, *A Web of Our Own Making: The Nature of Digital Formation* (New York: Cambridge University Press, 2023), 15.

Which Dystopia?

Just as today we look back to Postman's book to help make sense of our cultural moment, so too did Postman look to the past from his vantage point in 1985, at the peak of what he called the "Age of Show Business." The old books Postman looked to for insight were a pair of dystopian novels: George Orwell's *1984* (published in 1949) and Aldous Huxley's *Brave New World* (1932). Working on his book in 1984, Postman pondered: Had Orwell's vision of that year come to fruition? Or was Huxley's dark vision of the future more accurate?

Postman concluded that Huxley's dystopia, not Orwell's, better predicted the shape Western society took in the latter half of the twentieth century. As he explained,

> Orwell feared those who would deprive us of information. Huxley feared those who would give us so much that we would be reduced to passivity and egoism. Orwell feared that the truth would be concealed from us. Huxley feared the truth would be drowned in a sea of irrelevance. Orwell feared we would become a captive culture. Huxley feared we would become a trivial culture, preoccupied with some equivalent of the feelies, the orgy porgy, and the centrifugal bumblepuppy. As Huxley remarked in *Brave New World Revisited*, the civil libertarians and rationalists who are ever on the alert to oppose tyranny "failed to take into account man's almost infinite appetite for distractions." In *1984*, Huxley added, people are controlled by inflicting pain. In *Brave New World*, they are controlled by inflicting pleasure.[5]

---

5 Neil Postman, *Amusing Ourselves to Death*, 20th anniversary ed. (1985; repr., New York: Penguin Books, 2005), xxi–xxii.

If Postman was astute in 1985 to observe the Huxleyan shape of our "trivial culture"—where opted-in distractions and diversions kept us numb and dumb—how much more accurate does his prophetic vision describe life in 2025?

When Postman wrote *Amusing Ourselves*, he had television mostly in view as the chief purveyor of trivial information that swept us away in a "sea of irrelevance." Forty years later, we still have TV—albeit hundreds more channels and a growing number of streaming TV platforms. But we also have YouTube, Facebook, TikTok, and other always-on pipelines of content, algorithmically designed to grab our attention and keep us watching and scrolling, eyes glued to screens.

"Amusing ourselves to death" is still a highly accurate descriptor of what mass media does to us. But now the dominant form it takes is scrolling. And while Postman, who died in 2003, never lived to see the way smartphones, streaming, and social media would transform the world, his wisdom and warnings ring out with potent relevance.

Just as Huxley helped Postman make sense of his world in 1985, Postman can help us make sense of ours.

Who Was Neil Postman?

I first came across Postman when I was an undergraduate communications major at Wheaton College. One of my professors had studied with Postman at New York University. In helping us become better thinkers about the *forms* of media rather than just their *content*, he introduced my class to the concept of *media ecology*, which Postman had first coined, building on the work of Marshall McLuhan. I quickly devoured all the Postman books I could get my hands on, from *The Disappearance of Childhood* (1982) to

Technopoly (1992) to *Building a Bridge to the 18th Century* (1999), and, of course, *Amusing Ourselves to Death*.

This was in the early 2000s at the dawn of the internet age. I sensed the wisdom of Postman and McLuhan would be vital in my life as a Christian navigating a world of rapidly changing technologies. My interest in media ecology—particularly its implications for theology and the Christian life—led me to enroll in a media studies master's program at UCLA. Postman is in the backdrop of much of my writing about culture and the church.[6] I'm convinced he's a thinker whose wisdom is vital for the contemporary church. This present volume is an attempt to introduce Postman to a broader audience of Christians or to help those already familiar with him to apply his insights in helpful ways in their lives and ministry contexts.

Born into a Yiddish-speaking family in Brooklyn the year before Huxley published *Brave New World*, Postman lived in New York City for most of his life and became one of America's most prominent public intellectuals in the latter half of the twentieth century. He founded NYU's Steinhardt School of Education's program in media ecology in 1971 and was chair of the Department of Culture and Communication until 2002. Most of his work explored how media and technology influenced education, childhood development, politics, and public discourse. But he was also interested in how the "medium" influenced the "message" of religion and ostensibly sacred texts.

Postman wasn't a Christian. He was Jewish. But his faith informed his perspectives. *Amusing Ourselves* is full of references

---

6 See especially *Hipster Christianity: When Church and Cool Collide* (Grand Rapids, MI: Baker, 2010) and *The Wisdom Pyramid: Feeding Your Soul in a Post-Truth World* (Wheaton, IL: Crossway, 2021).

to God and the history of religious discourse in America going back to the Great Awakenings of Jonathan Edwards and George Whitefield's day. There's a clear sense that, of all the things that had become "trivialized" in the age of show business, Postman was most uneasy with how God and theology were being refashioned in the image of a TV variety show (televangelism was at its peak influence when Postman was writing).

The context in which Postman wrote *Amusing Ourselves* is important. The mid-1980s was the era of Ronald Reagan's presidency—a time when the highest office in the land was occupied by a former Hollywood actor (a point we hardly shrug at today but that was novel and ghastly to intellectuals like Postman at the time). The 1980s was also a period of TV's rapid expansion from three main networks to—with the onset of cable—dozens of channels, including the first-ever twenty-four-hour news network (CNN, in 1980).

In Postman's view, TV accelerated a seismic shift in the dynamics of information. For much of human history, he observed, we suffered from *information scarcity*. But now we have the opposite problem: *information satiation*. The information glut has many side effects, which Postman details in *Amusing Ourselves*. These include information trivialization in a "Now . . . this" flow of discombobulating coverage ("How serious can a flood in Mexico be, or an earthquake in Japan, if it is preceded by a Calvin Klein jeans commercial, and followed by a yogurt commercial?"[7]); a tendency toward impatience, forgetfulness, and poor logic in how we process information; and a massive shift in the formula for political success.

On politics Postman was especially prophetic. He observed a change in how voters picked leaders—no longer chiefly on the

---

7   "Life and Career of Neil Postman," *C-SPAN*, January 14, 1988, video, https://www.c-span.org/.

grounds of agreeing with (let alone understanding) the candidate's policies, but instead on the personality of the candidate ("Do I *like* this person?"). Image, branding, and "relatability" replaced issues. "We may have reached the point," Postman argued, "where cosmetics has replaced ideology as the field of expertise over which a politician must have competent control."[8] Keep in mind, Postman was observing this eight years before Bill Clinton played the saxophone on *Arsenio Hall* and told voters, in a town hall debate, "I feel your pain." He was seeing this trajectory three decades before the reality TV star Donald Trump became president.

On these and many other points, *Amusing Ourselves* was utterly prescient.

Applying Postman, Forty Years Later

Postman's critique isn't perfect. At times, he pushed too hard the idea that printed words are the only valid means of communicating important truths and fostering meaningful discourse. He occasionally comes across snobbish when, for example, he scoffed at the idea that an actor could become an effective US president. And while his discussions of American history are fascinating, the book at times feels too nostalgic for bygone eras in US history. Indeed, I wish Postman didn't single out American culture as much as he did. However much the United States tends to exemplify the media dynamics he critiqued, the problems he identified are everywhere, even more now than in the 1980s. In his *New York Times* review in 1985, Anatole Paul Broyard put it well: "Much of 'Amusing Ourselves to Death' is true, but it's not the whole truth and nothing but the truth."[9]

---

8   Postman, *Amusing Ourselves*, 4.

9   Anatole Broyard, "Going Down the Tube," *New York Times*, November 24, 1985, 9, https://www.nytimes.com.

Still, despite its flaws and its admittedly dated focus on analog television, the principles Postman offered are highly applicable today. Some of his insights have become even *more* incisive today than they were forty years ago.

One of Postman's key points about television as a medium, for example, is even more astute and important when applied to today's internet-shaped world. Postman argued that above all else, television's function is to gather an audience that can be sold to advertisers. TV exists as an efficient instrument for the advancement of corporate profits by delivering huge audiences of captive eyeballs. Here's how Postman put it in a C-SPAN interview in 1988: "In the past, audiences were gathered for specific reasons—to hear speeches or even to see specific events—but television doesn't do that. Its job is to gather an audience, and it doesn't really much care what it uses as the means to gather an audience."[10]

Postman argued that American television in particular discovered quickly that the best way to gather an audience was not to *responsibly inform* or *truthfully report* but to *constantly amuse*. Once we recognize this fact—that television is fundamentally oriented around commandeering your attention so it can be monetized—we can begin to resist its pull.[11]

The same is true today, even as the stakes are higher. The internet, like television, traffics in the currency of attention.[12] Every app, every website, every social media influencer whose bottom line depends on keeping eyeballs engaged is in the business of audience building. And the business of audience building—whether

10 "Life and Career."

11 See Shoshana Zuboff, *The Age of Surveillance Capitalism: The Fight for a Human Future at the New Frontier of Power* (New York: Public Affairs, 2019).

12 See Tim Wu, *The Attention Merchants: The Epic Scramble to Get Inside Our Heads* (New York: Knopf, 2016).

you're the *New York Times* or the NFL, Barack Obama or Blippi—is ultimately an *amusement* business.

What will captivate a scrolling eye long enough to pause the fidgety finger and get it to click? What content will maximally trigger adrenaline or dopamine rushes and cultivate addictive behavior, keeping audiences tuned in? Make no mistake: these questions drive almost every corporation, advertiser, editor, performer, creator, thinker, and influencer vying for attention in a vastly crowded media environment. And it has serious implications for them. And for you.

Goal of This Book

If the dynamics of the television age posed provocative questions for Christians in the 1980s, the dynamics of the internet age have only amplified the questions—and introduced many new ones—in the 2020s. For our own spiritual health, and to maintain a prophetic power and witness in a world being changed faster than it can even recognize, Christians in this cultural moment should slow down and think wisely about the ever-changing technologies swirling around us.

Sadly, many Christians default toward a naïve embrace of technology as a neutral and merely pragmatic tool to be harnessed for mission. But as Postman—following McLuhan—rightly argued, no technology is neutral. New technologies shape our thinking: what we think *about*, the symbols and metaphors we think *with*, and the forums in which thoughts develop.[13]

Technology's power to shape thinking should matter to every sensible person, but it should especially matter for Christians.

---

13 Neil Postman, *Technopoly: The Surrender of Culture to Technology* (New York: Vintage Books, 1992), 20.

After all, our mission revolves around transactions of thought: the gospel message being heard, understood, internalized, and applied. Because getting people to think well about God and the Bible (theology) is central to Christian mission, we must be aware of how thinking is changing as a result of different technologies and how our habits of worship, preaching, evangelism, and apologetics might need to adjust to these shifting dynamics.

This book is intended to get you thinking about technology, in part by showing you how current technology is changing the way we think. Using Postman's *Amusing Ourselves* as a jumping-off point, the contributors will explore various questions, challenges, and opportunities the church must grapple with in this highly formative technological moment.

The chapters in part 1 ("Postman's Insights, Then and Now") introduce some of Postman's core arguments in *Amusing Ourselves*, especially in light of what has changed since he wrote the book. The chapters in part 2 ("Practical Challenges Facing Christian Communicators") apply Postman's concepts particularly to the challenges facing gospel communicators (preachers, teachers, apologists, evangelists) in our contemporary context. Finally, the chapters in part 3 ("How the Church Can Be Life in a 'Scrolling to Death' World") turn from negative challenges to positive opportunities, suggesting ways the church can be a radical, life-giving alternative to the unhealthy habits of the digital world. If you feel a bit depressed reading some of the sober assessments in the first two sections of the book, hang in there and keep reading. The concluding chapters offer some positive visions—and practical recommendations—that give me hope.

This book makes the case that as Christians seek to wisely navigate our present—and future—media environment, we would do well to hear and heed Postman's clarion call. We look to Postman not as an all-encompassing explainer of everything or an all-knowing guide for the future but as a provocative voice that prompts necessary thinking and constructive conversations—not just for the sake of our own scrolling souls but also for the sake of our lost neighbors. The church mustn't stand by as scores of people scroll their way into oblivion, distracting themselves to death and clicking their way to corruption. We must step in and speak truth that gives life, redirecting glazed-over eyes and lifting hunched-down faces to behold the one who is infinitely more satisfying than whatever fleeting amusements flash across our screens.

Discussion Questions

1. Have you read Neil Postman's *Amusing Ourselves to Death*? If not, you should. It's a good companion to this volume. If you've read it, what stands out as the most prophetic insight from Postman as it relates to the technological world forty years after he published the book?

2. McCracken uses the phrase "currency of attention" and argues much of the internet is in the business of audience building, which "is ultimately an *amusement* business." What does it mean that your attention is so profitable for corporations and content creators online? How does this inform our understanding of scrolling habits as a matter of our spiritual formation?

3. Can you think of an example of a technology Christians uncritically adopted for a pragmatic purpose that eventually led to unforeseen negative consequences? Or a technology in our broader society that was introduced to solve a particular problem but ended up creating new problems?

4. What do you see as the most acute pain points for Christians and churches as they interact with new technologies? On what specific topics do we need to pursue the sort of "necessary thinking and constructive conversations" McCracken says this book is designed to spark?

PART 1

POSTMAN'S INSIGHTS,
THEN AND NOW

1

From Amusement to Addiction

Introducing Dopamine Media

Patrick Miller

IN 2011, Julijonas Urbonas unveiled a miniature model of his "Euthanasia Coaster." If built, the full-size roller coaster would be four-and-a-half miles long, beginning with a massive drop, followed by seven consecutive tightening loops accelerating to a lethal 10 Gs of force.

"It's a euthanasia machine in the form of a roller coaster," Urbonas explained, "engineered to humanely, with euphoria and pleasure, kill a human being."[1] The acceleration causes the rider to suddenly suffocate, inducing a brief, painless, euphoric state generated when the brain focuses only on vital activities.

---

1 Nate Swanner, "The 'Euthanasia Coaster' Was Designed to Kill Riders with Elegant Violence," The Manual, October 12, 2023, https://www.themanual.com/.

Euthanasia is unethical, so we can be thankful no such amusement exists. Nonetheless, you can find countless articles, YouTube videos, Reddit threads, and social media posts in which ordinary people share how much they like the idea. *Why not amuse yourself to death?* After all, it's strangely poetic for humans addicted to amusement to die by it. From entertainment you were made, and to entertainment you shall return.

No one likes to think of himself as an entertainment addict, wasting away his life on impulsively foolish, self-indulgent, self-destructive endeavors. But if an objective observer from a pre-digital era followed you around for a day and watched you compulsively check your phone, refresh your email, ogle at social media, binge videos, and tune out your children with AirPods, what conclusions would he draw?

Would he see an addict? Someone on a decades-long Euthanasia Coaster, slowly amusing himself to death in the way Postman predicted?

Postman's insights four decades ago can help us in our own time as we consider technology's trade-offs and how we're being shaped by our internet-era media environment, for good and ill. The internet, social media, mobile computing, and artificial intelligence have brought benefits we don't want to give away, but they've also come with costs. Is the trade-off worth it?

Postman understood that all "new technology for thinking involves a trade-off": "It giveth and taketh away, although not quite in equal measure. Media change does not necessarily result in equilibrium. It sometimes creates more than it destroys. Sometimes, it is the other way around. We must be careful in praising or condemning because the future may hold surprises for us."[2]

---

2   Neil Postman, *Amusing Ourselves to Death*, 20th anniversary ed. (1985; repr., New York: Penguin Books, 2005), 29.

So let's embrace carefulness, and ask, What are the trade-offs of internet-era digital technologies?

How Media Changes How We Think

Postman didn't live to see our current iteration of digital technology, but he modeled *how* to think through trade-offs—particularly how media changes the way we *think*.

In *Technopoly*, Postman reflected on a myth from Plato's *Phaedrus*, in which an Egyptian king named Thamus converses with the divine progenitor of reading and writing, Theuth. The god explains all the benefits which will accrue to humans who adopt his new media format. But King Thamus demurs. The trade-off isn't worth it: "Those who acquire [writing] will cease to exercise their memory and become forgetful. . . . They will rely on writing to bring things to their remembrance by external signs instead of by their own internal resources."[3]

Postman agreed with King Thamus that there were mental trade-offs with the widespread adoption of reading and writing. Memory was one. However, the written word also generated new mental worlds: theology, the natural sciences, ecology, economics, mathematics, philosophy, sociology, medicine, and much more besides.

The trade-off was real but worth it.

That's not the case with later technologies, Postman believed. He thought TV changed how we think for the worse. To be clear, Postman said he was *not* claiming "that changes in media bring about changes in the structures of people's minds."[4] But much of his work gives that impression. He argued that people in the

---

3 Plato quoted in Postman, *Technopoly: The Surrender of Culture to Technology* (New York: Vintage Books, 1992), 10–11.

4 Postman, *Amusing Ourselves*, 27.

televisual era learned and thought about truth differently than those in the prior (typographical) era.[5] Their televisual minds lacked the attention necessary to engage in the long-form discourse common in the typographic era.[6] Worse, the televisual mind lacked typographical fluency in abstract reasoning, preferring the concrete and emotive. Postman thought TV cultivated a highly subjective, highly expressive, highly therapeutic, and highly individualized way of perceiving the world and self, and of evaluating truth.

Those weren't merely changes to the structure of discourse. They were changes in the mind.

The same is true today. Yes, digital media changes the structure of discourse. But that's not all it changes. If Postman were alive, I suspect he might be nostalgic for the TV era. Because what followed TV is quite literally rewiring our brains.[7]

We're amusing ourselves into addiction. Entertainment culture metastasized into something not even Postman could have predicted: dopamine media. In some ways, the dystopia that inspired his work, Huxley's *Brave New World*, did see it coming. In that universe, a drug called Soma is used to anesthetize the people and keep them happy. Our addictive (more on this later) drug of choice isn't ingested or injected. It's consumed ocularly.

The trade-off we all make in the digital era is not merely between substantive and trivial discourse. It's between sobriety and

---

5 Postman, *Amusing Ourselves*, 61, 71.

6 Postman, *Amusing Ourselves*, 45.

7 The effects of this rewiring are most pronounced among children going through puberty. The long-term addictive effects described later in this chapter will only escalate with the aging of Gen Z, the first generation to receive *digital dopamine* (more on that phrase later). See Jonathan Haidt, *The Anxious Generation: How the Great Rewiring of Childhood Is Causing an Epidemic of Mental Illness* (New York: Penguin, 2024), 136.

addiction. While TV addicts have existed since television's inception, the technology wasn't addictive enough or constantly accessible enough to become dependence-forming. It's easy to think smartphones are just an extension of TV technology, but even though the phone in your pocket looks like a tiny TV, it's actually something far more nefarious.

Your phone is a digital syringe.

It's a gateway to lifelong, brain-altering, relationship-destroying addiction.

Digital Dopamine Nation

In *Dopamine Nation*, Stanford professor of psychiatry and behavioral sciences Anna Lembke argues pervasive, cheap, and easy-to-access products and experiences that release dopamine in the brain are creating a mental health crisis unlike any other in human history. This is for the simple reason that most people in history lived with scarcity—limited access to the foods, substances, and experiences that release dopamine in the brain—but now we live in a world of abundance. Our brains were not designed to live in such a world.

The consequence of dopamine abundance is addiction. To understand how this works, Lembke says it's helpful to imagine your brain like a seesaw. On one side is pleasure; on the other side is pain.[8] Your brain wants to retain equilibrium, to keep the seesaw flat.

The longer you spend with your mental seesaw tipped to pleasure, the harder the pain comedown. While your reflexive self-regulation mechanisms press the pain side down, you may

8   Anna Lembke, *Dopamine Nation: Finding Balance in the Age of Indulgence* (New York: Dutton, 2021), 51–53.

experience heightened levels of stress, depression, and irritability, and a whole array of psychological symptoms that make your brain want *more* dopamine to relieve your psychological distress.

Throughout most of history, it was hard to find substances and experiences that could press the pleasure side, so equilibrium was more commonly attained. But when you live in a society awash with dopamine factories—social media, pornography, gaming, high-calorie foods, alcohol, online gambling—you face a constant, pathological temptation to press the seesaw on the pleasure side.

The problem is that the more you repeat a dopamine-releasing behavior, the greater your tolerance becomes. This applies to social media—a proven dopamine-releasing substance—which was designed to be addictive.[9] Thus, if it took only two TikToks to spike your dopamine the first time, it will take four the tenth time, and dozens the hundredth. Whatever your drug of choice, you need more and more of it to get the original high *and* more and more of it to reduce the psychological pain you experience when you come down from your high.

It's a vicious cycle. Anyone who experiences ghost vibrations in his pocket—beckoning him to clutch his phone—knows this cycle. Anyone who's opened YouTube or Instagram to watch a video for five minutes only to inexplicably lose an hour knows this cycle. Anyone who cannot resist the impulse to watch digital pornography or gamble online knows this cycle. If the faintest shadow of boredom makes you compulsively check your phone, then you know this cycle. If you are easily distracted during a conversation with your spouse by the strange and desperate urge to check your

9 Lembke, *Dopamine Nation*, 191.

phone, then you know this cycle. If a brief moment of anxiety makes you swipe madly through your phone looking for *any* unread notification, then you know this cycle.

Your brain is seeking dopamine. It's whispering, "Get out the digital syringe. Take another hit. Then the boredom, stress, irritability, and blues will go away."

In the brain, what goes up must come down. And the comedowns from consistent use of dopamine media are causing a social and mental health catastrophe on a scale never before seen.[10] NYU psychology professor Jonathan Haidt analyzed countless studies to determine that social media and smartphones are causing this catastrophe, especially among our children. Teenage boys and girls are experiencing higher levels of depression, anxiety, and suicidal ideation.[11] According to a recent US surgeon general's advisory report, it's all correlated to smartphone use.[12] The advisory report urges parents not to give their children access to social media. Despite laws prohibiting social media usage under the age of thirteen without parental permission, 38 percent of children between the ages of eight to twelve are using regularly—many for hours a day.[13] Ninety-five percent of teens between thirteen and seventeen are using digital dope, and most parents can't bring themselves to tell them to stop, even though social media's dangerous and addictive effects are now widely known.

10 Haidt, *The Anxious Generation*, 14.

11 Jonathan Haidt, "Why the Past 10 Years of American Life Have Been Uniquely Stupid," *Atlantic*, April 11, 2022, https://www.theatlantic.com.

12 *Social Media and Youth Mental Health: The U.S. Surgeon General's Advisory*, US Department of Health and Human Services, 2023, https://www.hhs.gov/sites/default/files/sg-youth-mental-health-social-media-advisory.pdf.

13 V. Rideout, A. Peebles, S. Mann, and M. B. Robb, "Common Sense Census: Media Use by Tweens and Teens," (San Francisco, CA: Common Sense), 2022, https://www.commonsensemedia.org/sites/default/files/research/report/8-18-census-integrated-report-final-web_0.pdf.

The transition from entertainment culture to dopamine media culture created more addiction in more households. To resist this addiction, we must first understand what it is and how it addicts users. Only then can we explore pathways forward for Christians and churches.

Dopamine Media Is a Digital Las Vegas

Postman suggested every era in American history is represented by a city.[14] Boston was the apotheosis of revolutionary fervor. Chicago was the incarnation of industrial dynamism. New York was the personification of melting-pot America. And finally, Las Vegas became the avatar of overentertained America.

Postman was right about Las Vegas. The city is world-renowned for its extravagant, ubiquitous entertainment. But Vegas is more renowned for something else: gambling. And thus, it's also the ideal embodiment of the current phase of American history: dopamine media.

While most Americans tend to think of substances as addictive—especially those that directly deliver dopamine—new research shows that behaviors can be profoundly addictive as well because they release dopamine in the brain. In 2013, pathological gambling was reclassified as an addictive disorder by the *Diagnostic and Statistical Manual of Mental Disorders*. And the way gambling works on the brain is exactly how dopamine media works. Lembke explains: "Studies indicate that dopamine release as a result of gambling links to the *unpredictability of the reward delivery*, as much as to the final (often monetary) reward itself. The motivation to gamble is based largely on the inability to predict the reward occurrence, rather than on financial gain."[15]

---

14 Postman, *Amusing Ourselves*, 3.
15 Lembke, *Dopamine Nation*, 61. Emphasis added.

A 2010 study found those addicted to gambling experience higher levels of dopamine release not when they *won* money but when they stood an equal chance of winning or losing money.[16] The best dopamine high came from uncertainty, not victory. In other words, when it comes to dopamine, anticipation of a reward can create more pleasure than the reward itself.[17] A slot machine is addictive because it keeps you in an anticipation loop: the big win is always just around the corner, so you pull the lever one more time, releasing anticipation dopamine in your brain.

This insight is key because it's central to how dopamine media works. Behavioral psychologists in virtually every big tech corporation design their platforms and apps (social media, news media, video media) using intermittent variable rewards, what have been called digital slot machines. Natasha Schull, author of *Addiction by Design*—a book researching actual slot machines—explains that "Facebook, Twitter, and other companies use methods similar to the gambling industry to keep users on their sites."[18]

Every time you post on social media, you pull a digital lever and receive an intermittent variable reward. Sometimes you win two likes, sometimes you win two hundred. If you're scrolling through reels, some videos are duds but some make you squeal with laughter. The great appeal of short-form video content—pioneered by TikTok and replicated by Meta and YouTube—is that the brevity allows the user to pull the lever constantly. The brain is constantly

---

16 Jakob Linnet, Arne Møller, Ericka Peterson, Albert Gjedde, and Doris Doudet, "Dopamine Release in Ventral Striatum During Iowa Gambling Task Performance Is Associated with Increased Excitement Levels in Pathological Gambling," *Addiction* 106, no. 2 (February 2011), 383–90, https://doi.org/10.1111/j.1360-0443.2010.03126.x.

17 Lembke, *Dopamine Nation*, 62.

18 Mattha Busby, "Social Media Copies Gambling Methods 'to Create Psychological Cravings,'" *Guardian*, May 8, 2018, https://www.theguardian.com/us.

releasing dopamine as it anticipates a reward. When you lose and get a lame video, you experience brief frustration or boredom, which only sends you back for more.

Swipe. Swipe. Swipe. Swipe.

Every time we do it, we're rewiring our brains the same way gambling addicts do.

What sets dopamine media apart from entertainment media isn't just its slot machine design, however; it's dopamine media's constant accessibility and algorithmic curation.

In Postman's day, humans had limited access to TV. Physically, it was stationary. To watch TV you had to sit in a room with a large device that needed to be plugged in. Additionally, you could watch only what was being broadcast on certain channels at certain times, on a schedule you didn't design. While cable networks tried to curate more niche-based spaces—think HGTV, the Food Network, or Comedy Central—television was never actually personalized.

Dopamine media is entirely different. It is physically unencumbered, traveling on your person and accessible anywhere. It's also temporally unconstrained. There aren't schedules. Thus, you can access *whatever* media you want, *whenever* you want, *wherever* you want.

But here is the real secret sauce: artificial intelligence. Everything you see on virtually every app and platform—from ads to videos to posts to search results—is generated by recommender algorithms: advanced AIs that use your data to create a digital model of you so it can feed you bespoke content to keep and monetize your attention.[19] Your social media feed is bespoke. It is designed to keep you specifically addicted, by AIs whose computational knowledge of you is shockingly vast and actionable. Their main job is to keep

---

19 Lev Grossman, "How Computers Know What We Want—Before We Do," *Time*, May 27, 2010, https://time.com/.

you on the platform—to keep you addicted—by tracking your behavior like a dystopian digital Pavlov.

Let's try to bring all this together in a chart highlighting how different today's dopamine media ecosystem is from the TV-entertainment ecosystem of Postman's day.

	Entertainment Media	Dopamine Media
Physical Access	Limited by large TVs and plugs	Unlimited; available anywhere on mobile devices
Temporal Access	Limited by TV schedules	Unlimited; available anytime, on-demand
Personalization	Directed toward large audiences based on broad viewing data	Calibrated for individuals based on their personal data
Curation	Content curated by humans to resonate with broad audiences	Content curated by advanced AIs to addict particular individuals
Length	Programs run thirty to sixty minutes	Micro content: thirty to ninety seconds, sometimes shorter
Variable Rewards	Limited by what was available via channel surfing	Constant; digital slot machines

"Amuse" doesn't quite describe the effect dopamine media has on us. Dopamine media is designed to *distract* us to death. Or, if we're more honest, to distract us into an addiction that leads to death. Research shows that the more available and normalized a drug is, the more pervasive addiction to that drug becomes. So it's no surprise the vast majority of American adults are walking around shooting up digital dope without raising an eyebrow. The best of us are responsible users who can consume media in moderation. But none of us is fully sober.

The addiction trade-off that dopamine media offered us isn't a *possibility*; it's already here. And if the first victims of our addiction are our time and attention span, the second (and far more important) victims are our families and relationships.

Research shows that the more addicted you become to dopamine-producing behaviors, the less your brain rewards you for being in relationship with others. This is even true of rats: if a free rat finds a caged rat, it will try to free it. But if you allow that rat to self-administer heroin, it will no longer be interested in the caged rat. The heroin gives a better high, after all.

Our addiction to dopamine media is training us to love much what ought to be loved little. It's making us miserably unhappy, hurting our relationships, and demanding more and more of our time to get the next high. Augustine wrote,

> The person who lives a just and holy life is one who is a sound judge of these things. He is also a person who has ordered his love, so that he does not love what it is wrong to love, or fail to love what should be loved, or love too much what should be loved less (or love too little what should be loved more), or love two things equally if one of them should be loved either less or more than the other, or love things either more or less if they should be loved equally.[20]

Dopamine media is the most powerful, pervasive, and engineered form of communication technology in human history, and it's not shaping us to love Jesus most. It's not shaping us to love our neighbor. It's shaping us into pleasure-seeking addicts. Christians must recognize that, at its heart, this technological revolution has resulted in an institutional, relational, and formational crisis for the church.

20 Saint Augustine, *On Christian Teaching*, trans. R. P. H. Green, Oxford World's Classics (Oxford: Oxford University Press, 1997), 1.27–28.

Institutional Crisis for Churches in Digital Las Vegas

Yuval Levin defines institutions as "the durable forms of our common life" and "the frameworks and structures of what we do together."[21] While I don't expect most people to get excited by words like *institution* and *institutional*, Christians must understand that Jesus not only announced "the gospel of the kingdom" but also established that kingdom by his death and resurrection. His kingdom is, of course, a durable social structure that orders common life and gives a framework not only for our ethical norms but also for the smaller structures (families, small groups, communes) that collectively form the larger ones (churches, parishes, denominations). Local churches are designed to bridge God's kingdom on earth and heaven.

Brad Edwards, a pastor and writer, argues that social media platforms are "pseudo-institutions" and "counter-institutions."[22] They mimic what real-world institutions can offer—think faux community, faux discourse, faux authenticity, faux intimacy, faux mentorship, faux wisdom—and in the process destabilize the very institutions they mimic. As digital addiction drives more people to seek influence and mentorship online, localized institutions will undergo a crisis of authority. As digital addiction drives more people to find connections and conversation online, they will undergo a crisis of community. As digital addiction drives more people online to find information and wisdom, they will undergo a crisis of moral norms.

What does this mean for the future of the church and evangelism globally? At the very least, it means Christians must

---

21 Yuval Levin, *A Time to Build: From Family and Community to Congress and the Campus, How Recommitting to Our Institutions Can Revive the American Dream* (New York: Basic, 2020), 19.

22 Brad Edwards, "The Church amongst the Counter-Institutions," Mere Orthodoxy, April 1, 2021, https://mereorthodoxy.com/.

recontextualize the gospel not only in light of their local milieu but in light of a global digital milieu made up of hundreds of thousands of AI-tailored microcultures. This is no small task, and it's one that requires a sovereign, transcendent, all-knowing Lord to guide us.

Thankfully, such a person sits on the throne of heaven.

As much as we may wish to cloister ourselves from dopamine media, we must instead take confidence in the power of God's grace. He knows more than the AIs. He has more resources than big tech. His spirit can heal broken minds. He commands time itself. We're not on the losing side of a pointless battle. Instead, we're serving a King who's calling us to ask once more how we can be faithful in our generation and offer his healing in a broken, digital dopamine-addicted world.

Discussion Questions

1. Is Miller right in classifying our phones with other drugs? Consider how many times you pick up your phone or computer each day. If your phone is a "digital syringe," would you say you're a tech addict?

2. Now that most of us have an idea of the drastically negative mental health effects of "big tech," what prevents us from taking action on this knowledge? How can the church help communities and individuals break free from this cycle?

3. After reading this chapter, try listing out some of the costs and benefits of the internet, social media, and artificial intelligence. Why is it so critical to consider whether a technology is a net gain or net loss for us as individuals and communities?

4. At the end of the chapter, Miller posits that the advent of do-pamine media has resulted in a "crisis for the church." How might dopamine media directly oppose the Great Commandment (Matt. 22:36–40)?

2

From the Clock to the Smartphone

A Brief History of Belief-Changing Technologies

Joe Carter

SCI-FI NOVELIST William Gibson said, "The future is already here—it's just not evenly distributed."[1] Gibson meant that what will be normal in the future already exists for some people today. Take, for example, the internet. The birth of the internet is considered to be January 1, 1983, which means it was two years old when Neil Postman published *Amusing Ourselves*.[2]

The media critic had no idea of the future that was already there, so he wasn't able to speak directly about how the internet and other innovations would affect the world. But he didn't need to; he had already warned us.

---

1   "Broadband Blues," *Economist*, June 21, 2001, https://www.economist.com/.
2   Caitlin McLean, "When Was the Internet Invented? What to Know about the Creators of It and More," *USA Today*, August 28, 2022, https://www.usatoday.com/.

Postman's central argument—that technology is not neutral but rather a medium with inherent biases that can unintentionally shape our perceptions and values—is truer now, in the age of the internet, than when he penned those words.[3] This is especially evident in the realm of Christian belief, where the adoption of new technologies has often been driven by a pragmatic desire to harness their potential for spreading the gospel and supporting ministry. To propagate the good news, we've eagerly embraced innovation from the printing press to radio broadcasts to smartphone apps.

But in our enthusiasm, we haven't always paused to consider the unintended spiritual consequences. Could the tools we deploy for evangelism and discipleship be eroding the foundations of our faith? Is it possible that in our quest for relevance and reach, we've unwittingly altered the very plausibility structures that undergird Christian conviction?

By tracing the subtle but significant ways various technologies have reshaped theological understanding, I hope to provide a framework for discerning both the opportunities and challenges of our current digital age.

Is It Plausible?

Postman's critique invites us to consider the unintended consequences of technology and its potential to accelerate secularization. In particular, I want to consider a possible mechanism for how such secularization occurs: disruptive technologies change and shape our theological plausibility structures.

We tend to think we believe something merely because it's true. But all beliefs are shaped by plausibility structures—cognitive

---

3   Neil Postman, *Amusing Ourselves to Death*, 20th anniversary ed. (1985; repr., New York: Penguin Books, 2005), 22–23.

frameworks that act as gatekeepers, filtering information based on what we already consider possible.[4] These structures determine whether a claim seems reasonable or potentially true, without necessarily confirming its veracity. They allow us to accept claims that appear plausible enough and provide a basis for thinking they *could* be true. Our worldviews, whether gullible or skeptical, are formed by the accumulation of beliefs that have passed through these plausibility structures at both individual and cultural levels. Over time, our worldviews become broad filters that automatically reject beliefs we won't even consider as *possibly* true.

Imagine, for instance, that you come across a news article claiming a celebrity has been secretly replaced by an identical clone. The article presents "evidence" such as subtle changes in the celebrity's appearance or behavior. However, your plausibility structure, shaped by your understanding of human biology and the complexities of cloning technology, makes it challenging for you to accept this claim. You don't consider it possible, so you don't believe it to be true. The same process works for all our beliefs.

Plausibility structures aren't just personal intellectual constructs, though. They encompass a wide range of elements including cultural, social, psychological, and historical factors that make a certain set of theological ideas seem plausible or implausible to a person or

---

4 The term *plausibility structure* was coined by Peter Berger, who added, "Every human society is an enterprise of world-building. Religion occupies a distinctive place in this enterprise. . . . Each world requires a social 'base' for its continuing existence as a world that is real to actual human beings. This 'base' may be called its plausibility structure. . . . The reality of the Christian world depends upon the presence of social structures within which this reality is taken for granted and within which successive generations of individuals are socialized in such a way that this world will be real to *them*. When this plausibility structure loses its intactness or continuity, the Christian world begins to totter and its reality ceases to impose itself as self-evident truth." Peter L. Berger. *The Sacred Canopy: Elements of a Sociological Theory of Religion* (Garden City, NY: DoubleDay, 1969), 45.

a community. They play a crucial role in shaping how individuals and groups perceive, interpret, and respond to religious doctrines and experiences.

With this in mind, let's now reflect on how the clock, television, and smartphone have contributed to changing theological plausibility structures in ways that lead to the erosion of Christian belief and practice.

How the Clock Changed Our View of God

The mechanical clock—a seemingly innocuous invention—fundamentally altered human perception of time and divine providence. In his exploration of the clock's influence, Postman underscored its role in shaping a secular mindset.[5] Initially developed to regulate the schedule of Catholic monks, the mechanical clock soon transcended its clerical confines, introducing a new temporal consciousness to society.

The clock, by its very nature, abstracted time from the natural, God-ordained rhythms of the universe, laying the groundwork for a more secularized approach to life. The clock's mechanical representation of time challenged the traditional Christian understanding of time as cyclical, that our lives must center on the liturgical calendar and the agricultural seasons.

The mechanical clock thus heralded a subtle-yet-seismic shift in theological understanding that first began in Christian Europe. Prior to the clock's widespread adoption, time was perceived as a divine mystery, governed by the celestial bodies God set in motion. This celestial time was less about accuracy and more about a rhythmic harmony with creation.

5 Postman, *Amusing Ourselves*, 10–11.

However, as the clock began to dictate the cadence of daily life, it introduced a new conception of time as a quantifiable and controllable entity. This marked a departure from a God-centered universe to a human-centered one, where time became a resource to be managed and optimized, inadvertently nurturing a secular worldview. This change affected how we think about such topics as history, eschatology, and the nature of divine providence. Postman went so far as to say, "The inexorable ticking of the clock may have had more to do with the weakening of God's supremacy than all the treatises produced by the philosophers of the Enlightenment."[6]

The clock's intricate mechanisms and precise functioning also led to the idea of God as the ultimate clockmaker, creating the universe with perfect order and regularity due to fixed, deterministic laws. This metaphor of the "clockwork universe" became a dominant theological framework, suggesting God had set the world in motion and then stepped back, allowing it to run on its own. This challenged traditional notions of divine intervention and miracles, as it suggested God did not actively participate in the world's day-to-day operations, leading to a rise in Deism in the early seventeenth century.[7]

All these changes had far-reaching implications for religious thought and practice and contributed to the development of new theological frameworks that sought either to reconcile the clockwork universe with traditional Christian beliefs or to replace them entirely.

---

6   Postman, *Amusing Ourselves*, 11–12.
7   Alister McGrath, "The Clockwork God: Isaac Newton and the Mechanical Universe," Gresham College, January 23, 2018, https://www.gresham.ac.uk/sites/default/files/2018-01-23_AlisterMcGrath_TheClockworkGod.pdf.

Television's Secular Sermonizing

Postman argued that, like the mechanical clock, television has a "strong bias [as a medium] toward a psychology of secularism."[8] This assertion stems from the idea that television is ill-suited for conveying complex, abstract ideas inherent in religious doctrine. Instead, it favors immediate, emotive, and visually engaging content—making all these attributes more aligned with secular entertainment than with the nuanced and reflective nature of faith. One of the most powerful effects of television is the way it has transformed secular "sermonizing."

We evangelicals tend to think of sermons as a communication form restricted to the church. But as Kevin Simler points out, "mass moralizing" in secular contexts has the same effect as sermons.[9] Simler defines a sermon as "any message designed to change or reinforce what a group of people value." In other words, sermons function to change or reinforce what a group values by generating common knowledge and producing a network effect around certain ideals.

Effective secular sermons, as Simler argues, produce "common knowledge." As he explains,

> This means it's not sufficient for everyone in the audience to hear and [understand intuitively] a sermon *individually*, as it is with a lecture. Instead, for maximal effect, everyone has to know that everyone *else* has heard and [understood intuitively] the sermon as well. The more thoroughly this kind of knowledge saturates within a community—everyone

---

8 Postman, *Amusing Ourselves*, 119.
9 Kevin Simler, "Here Be Sermons," Melting Asphalt, September 11, 2017, https://melting asphalt.com/.

knowing that everyone else knows that everyone else knows that . . . everyone understands the sermon—the stronger the resulting network.[10]

In the past, these types of common knowledge-producing sermons were often explicit and delivered in front of a live audience, such as commencement addresses or presidential inauguration speeches. Television, however, made it possible for such sermons to be more subtle and delivered to a mass audience.

Consider, for example, one of the most effective (and expensive) forms of sermons: the Super Bowl ad. Due to the massive, simultaneous audience that tunes in for the game, Super Bowl ads provide prime examples of how secular sermons work to produce common knowledge. When a brand spends millions to place an ad during the Super Bowl, it's not just attempting to persuade individual viewers to have positive associations with the product (though it may be doing that too). More importantly, it's seeding what Simler calls a "cultural signal" that will be shared by millions of people watching.[11] The ad is a sermon, the message is the cultural signal, and the goal is to produce common knowledge.

Imagine Budweiser runs an ad portraying their beer as the ultimate party beverage. By airing this ad during the most-watched television event of the year, Budweiser ensures a huge segment of the population will absorb this cultural signal simultaneously. The next time a viewer goes to buy beer for a party, he'll choose Budweiser, not necessarily because he is personally imbued with good feelings about the beer but because he knows that *everyone*

10 Simler, "Here Be Sermons."
11 Kevin Simler, "Ads Don't Work That Way," Melting Asphalt, September 18, 2014, https://meltingasphalt.com/.

else saw that Super Bowl ad too. Bringing Budweiser will thus send the right cultural signal—that he's ready to party.

The Super Bowl ad will be more effective at cultural imprinting than if Budweiser had spent the same budget to reach an equal number of viewers separately, such as through targeted online ads. It's the mass simultaneous viewership and common knowledge that gives the Super Bowl ad its special power and what makes it a sermon. As Postman himself said, "Television commercials are a form of religious literature. . . . The majority of important television commercials take the form of religious parables organized around a coherent theology."[12]

Applying this framework of sermonizing more broadly, we can see how television can and has altered theological plausibility structures in society. TV shows and advertising that touch on religious themes act as virtual sermons preached to mass audiences. When television content portrays certain religious ideas or behaviors positively or negatively, it shapes common knowledge about the acceptability and popularity of those theological positions.

For example, a popular TV show that mocks religious belief or portrays faith as outdated and irrational makes skepticism towards religion seem like the default, common perspective. Animated series like *South Park* and *Family Guy*, for instance, have satirized Christian beliefs, practices, leaders, and denominations—sending the message that their audience should not want to be associated with these objects of mockery. The more entertaining and widely viewed the show, the greater the "sermon effect" in modifying plausibility structures.

12 Neil Postman, *Conscientious Objections: Stirring Up Trouble About Language, Technology and Education* (New York: Knopf Doubleday, 1992), 64–65.

But couldn't television be used to sermonize in a way that benefits the faith? Possibly. But Postman argued that television as a medium is biased toward entertaining content over serious discourse. Substantive arguments are simplified into sound bites and spectacle.

As we've seen throughout the medium's history, television's propensity to prioritize entertainment value over content depth has contributed to a cultural shift where religious ideas are increasingly judged by their ability to entertain and hold the audience's attention, rather than by their spiritual truth or theological depth. This shift has had profound implications for how religion is perceived and practiced. It has contributed to a form of "consumer" Christianity, where the success of a religious message is measured by its popularity and appeal more than its doctrinal soundness.

Applied to Christianity, this suggests TV has spread a superficial, feel-good, comedy-compatible version of faith rather than a weighty reverence for orthodox doctrine. The problem, as Postman would likely say, is that it's the nature of television to make a folksy emotive spirituality more plausible than a rigorous rational theology.

Disembodied and Deconstructing on the Internet

The advent and proliferation of the internet represented a watershed moment in technological history. The emergence of smartphones then added a new dimension that has transformed and multiplied the effect. By condensing the power of the internet into portable devices, smartphones have become a constant presence in most people's lives. They offer immediate access to information, entertainment, and, significantly for our discussion, disembodied social interaction.

The internet age presents a paradoxical situation for Christians. Never before has there been such easy access to biblical materials, teachings, and communities. Bible apps, religious-themed podcasts, and online sermons have made it possible to engage with faith-based content anywhere and anytime. Yet this accessibility comes with the risk of trivializing religious engagement, reducing it to just another form of digital content consumed passively and intermittently.

An even greater danger, at least for theological plausibility structures, is that it allows our engagement to be primarily disembodied. As Samuel James says,

> Because the Internet allows us to communicate with real language apart from our physical bodies and physical spaces, we tend to identify our "selfs" online not as whole-persons (body + mind) but as minds that exist independently of physical constraints. . . . Digital language tends to create a digital self-perception. And this digital self-perception makes certain theological claims less plausible than they might be otherwise.[13]

James gives the example of gender distinctions. Differences in gender make intuitive sense in contexts where we engage with one another as physical beings in each other's physical presence. But that changes, James argues, when our interactions are frequently, or even primarily, online:

> For the purposes of digital selfhood, the body is irrelevant. What creates personality online is language, language divorced from physical givenness. For a person who came of age experiencing the world through the Web, the idea that men and women might

13 Samuel D. James. "The Internet as a Theological Plausibility Structure," Digital Liturgies, Substack, May 15, 2023, https://www.digitalliturgies.net/.

be ordered toward differing patterns of life feels contrary not only to ideology, but to lived experience. It just doesn't make sense in an online world of bodiless personality to say that men and women are meaningfully different.[14]

Such disembodied exchanges have a powerful effect on what we know, whom we trust, and what we believe. As Tim Keller said, "Human knowledge has a (1) rational/intellectual aspect, a (2) experiential/intuitive aspect, and a (3) social/pragmatic aspect. That is, we come to 'know' something well when (1) there are good reasons for it, when (2) it fits with our inward experience, and when (3) we find a trustworthy community that holds it too."[15]

Internet-enabled devices change our knowledge by changing what we experience and who we consider trustworthy communities. For instance, the echo chambers often created by online algorithms can reinforce narrow worldviews, including secular or atheistic ones, potentially accelerating the process of secularization.

In recent years, the Christian community has witnessed an increasing trend toward deconstruction, a process where individuals critically analyze and often disassemble their inherited religious beliefs.[16] This movement, predominantly among younger generations, involves questioning and sometimes abandoning traditional doctrines and practices. While the causes of deconstruction are multifaceted, the rise of digital technology, particularly social media and the internet, has played a significant role in accelerating this phenomenon.

14 James, "The Internet."

15 Timothy Keller (@timkellernyc), X, February 28, 2022, 11:19 a.m., https://twitter.com/tim kellernyc/status/1498332140248961033.

16 Ivan Mesa, ed., *Before You Lose Your Faith: Deconstructing Doubt in the Church* (Austin, TX: The Gospel Coalition, 2021).

The digital age, marked by the widespread use of the internet and smartphones, has opened up an unprecedented avenue for the exchange of ideas and information. With just a few clicks, individuals can access a vast array of perspectives, including those critical of or even antagonistic to Christianity. They also find, through online forums and social media platforms, disembodied communal spaces where doubts and criticisms about faith are shared and amplified. Where in previous eras a doubting Christian might more easily move through and beyond her doubts in part because she couldn't easily find other believers who shared and could validate those doubts, today entire communities of every genre of doubt are just a Google search away.

The church faces a significant challenge in this technological ecosystem. Traditional modes of teaching and community engagement are being outpaced and sometimes undermined by the rapid flow of digital information and interaction. Young believers, or those in periods of doubt, often turn first to online sources rather than to church leaders or communities for guidance. The church must find ways to effectively engage with and minister to people in the digital spaces where they are increasingly living out their faith journeys.

Redeeming Technology in a Secular Age

From the mechanical clock to smartphones, every new technology brings both challenges and opportunities for the Christian faith. A prime example of an exciting—yet potentially new belief-killing—technology is artificial intelligence (AI). A series of studies suggests that as AI-enhanced automation increases, religiosity decreases. The researchers say, "There are meaningful properties of [AI-enhanced] automation which encourage religious decline."[17] Just reading about AI had a stronger negative effect on participants' religious convictions than reading about other kinds of breakthroughs.[18]

In our secular age, where technology often seems to erode spiritual depth and undermine embodied community, there is an urgent need for the church to proactively engage these issues. We must seek to redeem and reorient technology to serve kingdom work.

Rather than uncritically embracing the latest innovations or reacting against them, we need a framework for wisely discerning technology's effects—both positive and negative—on our faith and practice. This means asking hard questions about how a given technology shapes our perceptions, affections, and imaginations. It means considering its implications for spiritual formation, biblical literacy, relational intimacy, and more.

Ultimately, the call isn't to abandon technology but to bring it under Christ's lordship. We need to recognize that while technology will continue to change and evolve, "Jesus Christ is the same yesterday and today and forever" (Heb. 13:8). In him we have enduring hope, unshakable truth, and eternal life—the things we crave that our secular age and its technologies can't supply. By keeping our eyes fixed on Jesus, we can boldly and creatively use technology for his glory and for the good of his church.

Discussion Questions

1. How does Carter argue that the clock, the television, and the smartphone redefine our understanding of God? Do you find his arguments convincing?

---

17 Jeff Cockrell, "Where AI Thrives, Religion May Struggle," *Chicago Booth Review*, March 26, 2024, https://www.chicagobooth.edu/review/where-ai-thrives-religion-may-struggle.

18 Cockrell, "Where AI Thrives."

2. Carter points out that in the digital age we have unwittingly been taught to consider ourselves as "disembodied." After all, online, we don't have bodies; we're just minds. How might this belief affect the way we view gender, sexuality, and some of our world's hot topics today?

3. How is your phone—not the online content, but the device itself—training you to understand the world? How do the buttons, features, and digital layouts form Christians?

4. The success of a spiritual video, post, or message on the internet will be measured by likes, shares, reposts, and comments. Why might this be damaging to the church? Should Christians pay any attention to those sorts of metrics?

3

From the Age of Exposition
to the Age of Expression

Jen Pollock Michel

IN EARLY 2023, Dave Hollis was found dead with a phone on his chest. According to the *Wall Street Journal*, it was a "tragic" end to an "Instagram-perfect life."[1]

Dave's ex-wife, Rachel Hollis, had rocketed to stardom (at least temporarily) with the publication of her 2018 bestseller, *Girl, Wash Your Face*. The book and subsequent multimillion-dollar business capitalized on a glaring irony: that self-improvement was just as essential as imperfection.

Dave left his executive position at Disney to dive alongside Rachel into the bleeding world of indiscreet self-disclosure. Before

---

1 Quotations pertaining to the Hollis story are from Erich Schwartzel, "Behind the Tragic, Instagram-Perfect Life of an Ex-Disney Executive," *Wall Street Journal*, December 2, 2023, https://www.wsj.com/.

the couple's divorce in 2021, the two "mined their everyday lives" for content. Their brand of gritty "authenticity," performed for an audience, was effective. Yet private life could not support the weight of public exposure.

After the couple's divorce, Dave doubled down on the tell-all strategy. He wrote a self-help book. He continued the public live stream of his private life. But soon, not all that glittered on Instagram was gold. Hollis checked himself into rehab after admitting on social media that he was "feeling completely broken from the pressure of this strange public life."

When Hollis returned to social media four months later, his nearly half a million followers "saw an avatar of health online, spray-tanned and teeth whitened." His handful of next-door neighbors, by contrast, witnessed another Dave who returned to self-destructive addictions. After Hollis was found dead after an overdose, neighbor Lynn McKay blamed "viral social media fame and the platforms that power it" for his death.

"They're all drugs," she said.

From the Age of Exposition to the Age of Entertainment

Hollis is a tragic figure in a familiar contemporary drama. Authenticity is not just a personal brand in our digital age; it's a compelling (and often self-destructive) virtue. If, as Neil Postman argued, television conditioned our appetite for amusement, social media now conditions a craving for confessional intimacy. "Telling all"—from the privacy of your living room—has become big business, and it sells and rewards until, as in Hollis's case, it isolates and destroys. In our brave new world, self is a commodity and "reality" is entertainment.

The trouble began, Postman argued, with a shift away from print culture. His concern was our intellectual demise as a nation, as we left the "Age of Exposition" and arrived—with the dawn of television—in the "Age of Show Business." He worried less about television as a purveyor of content than as a tool of conversation: "What is television? What kinds of conversations does it permit? What are the intellectual tendencies it encourages? What sort of culture does it produce?"[2] Postman's preoccupation was *formation*. Ask not, in other words, what your tool does *for* you; ask what your tool makes *of* you.

According to Postman, the movement from word to image (and seriousness to silliness) was, at least initially, an American phenomenon. From its inception, America, as a national project, depended on the democratization of individual thought and expression. Perhaps we need look no further than the example Postman offered of Thomas Paine, the "unschooled shoemaker from England's impoverished class" whose 1776 book, *Common Sense*, skyrocketed to bestseller.[3] Only in colonial America, with its high literacy rates, its homegrown printing operations, and its independent sensibilities, was such democratic "voice" possible. Self-governance, in other words, demanded self-expression.

Yet self-expression in the abstract was not the problem Postman wished to identify in his very modernist critique. Rather, it was the loss of rational rigor that print culture purportedly imposed on self-expression when contrasted with television's discourse of fleeting images. In Postman's view, a citizenry that reads becomes serious. On a diet of television, a citizenry turns silly. Inevitably, in the age of show business, all content—political, religious, educational— must be performed. "Television does not extend or amplify literate

---

2 Neil Postman, *Amusing Ourselves to Death*, 20th anniversary ed. (1985; repr., New York: Penguin Books, 2005), 84.

3 Postman, *Amusing Ourselves*, 34–35.

culture," Postman argued. "It attacks it."[4] Television, he warned, "has made entertainment itself the natural format for the representation of all experience."[5] With television as our dominant media, Postman argued, we want to be titillated, not taught.

Individuating Revolution

Postman did not live to see the way life itself is "performed" on social media today, but his critique proved prescient. Pastors and teachers and politicians alike recognize the shortened attention span of their audiences, and they themselves feel the pressure to amuse. Still, it's not hard to admit that Postman was too enamored with the virtues of print. His defense of "semantic, paraphrasable, propositional content" and the "rational activity" of reading and writing ring with naive assumptions about written content and our unfettered capacity to apprehend truth.[6]

Written words, like images, share the capacity to distort truth. As one simple example, even the "truth-telling" of literary memoir, a contemporary genre of self-expression I studied in graduate school, is often questionable. Psychological research today emphasizes memory's unreliable nature. We can live an event and narrate it inaccurately. There is nothing infallible about words on a page.

But print bias isn't the only flaw in Postman's argument. Centering his account as a history of media, Postman neglects the larger cultural shift that sociologist Robert Bellah and his colleagues call the project of "expressive individualism."[7] Importantly, this drift—

---

4 Postman, *Amusing Ourselves*, 84.

5 Postman, *Amusing Ourselves*, 87.

6 Postman, *Amusing Ourselves*, 49.

7 Robert N. Bellah, Richard Madsen, William M. Sullivan, Ann Swidler, and Steven M. Tipton, *Habits of the Heart: Individualism and Commitment in American Life*, updated ed. (Berkeley: University of California Press, 1996), 27.

away from external sources of authority and toward the authority of the self—happens as the world changes due to industrialization and urbanization. People move, and crucial ties are severed with family and church and communal identity. This gives rise to expressive individualism, which might be distilled into today's cultural slogans: *You be you. Be true to yourself. Follow your heart. Find yourself.*[8]

In *A Secular Age*, philosopher Charles Taylor calls this the "individuating revolution." He means that we begin to prize a sense of "inner integrity"—a coherence between our inner desires and outward behaviors and commitments—over societal conformity.[9] Individual expression, then, requires some kind of renunciation of institutional authority, some small or large rejection of cultural propriety. This is the move of the American Revolution, and it is the move of "authenticity" as it's now posited in our age of expression. *How brave to be true to yourself!* When self-expression rises to prominence, authority becomes suspect and sin disappears. What matters is that human beings enjoy the freedom to decide for themselves.

In *Self-Made: Creating Our Identities from Da Vinci to the Kardashians*, Tara Isabella Burton attempts to make sense of today's call "to be self-creators: people who yearn not just to make ourselves a gift to the world but to make ourselves, period."[10] Burton espouses no naïve nostalgia about Postman's "Typographic America," and she's not even diametrically opposed to the place we've arrived. In Burton's view, self-expression has exposed much injustice, and to the degree that we can construct an identity beyond intersections of personal trauma and cultural oppression, all the better.

8 Trevin Wax, "Expressive Individualism: What Is It?" *Kingdom People* (The Gospel Coalition blog), October 16, 2018, https://www.thegospelcoalition.org/.

9 Charles Taylor, *A Secular Age* (Cambridge, MA: The Belknap Press, 2007), 474, 475.

10 Tara Isabella Burton, *Self-Made: Creating Our Identities from Da Vinci to the Kardashians* (New York: Hachette, 2023), 3.

Burton convincingly argues that self-expression, as virtue (and vice), has been underway since the European Renaissance. The seed—for tragic stories like that of Dave Hollis—was planted long ago, when we began to turn away from the anonymity of the medieval artisan and turn toward acts of self-creation and its necessary counterpart, self-promotion. Burton's history features figures like the sixteenth-century Renaissance painter Albrecht Dürer— "inventor of the selfie"[11]— who had a penchant for self-portraits in which he depicted himself as Jesus Christ.

By the time we arrive at the twentieth century, Burton notes self-will is fully divinized by the philosopher Friedrich Nietzche. God is dead—and the self reigns supreme. "A human being had two choices," Burton explains. "He could either slavishly allow himself to be determined by other people's values and self-assertions, or he could look inward, refashioning reality in accordance with his own self."[12] Nietzsche's formulation is the ripened fruit of an epistemology—a theory of knowledge, or how we come to know things—that had been changing over centuries. *How do we know what is true?* If the medieval man looked upward, the modern man, cured of such superstition, looked inward.

He consulted his *heart.*

Lose the Self—and Find It in Christ

In February 2024, *New York Times* columnist Pamela Paul called into question the aggressive tactics of modern medicine as related to gender dysphoria. Gender reassignment is, of course, one of the more obvious manifestations of "expressive individualism" today. Its philosophical assumption is that we are free, perhaps even obliged,

11 Burton, *Self-Made,* 12.
12 Burton, *Self-Made,* 138.

to construct an identity as determined by our inner sense of self. Our bodies don't define us; our desires do.

A self-avowed progressive, Paul is no regular defender of socially conservative positions. But in this case, she had fair and honest criticism for her more leftist friends. "Right-wing demagogues are not the only ones who have inflamed this debate," she writes. "Transgender activists have pushed their own ideological extremism, especially by pressing for a treatment orthodoxy that has faced increased scrutiny in recent years."[13] Despite the backlash she would inevitably face, Paul put her finger on one evident contradiction of the transgender movement. For as much as the freedom to *live and let live* has been flown as its ideological banner, the agenda seems quickly to have become dogmatic, intolerant, even doctrinaire.[14]

This is just one contradiction we meet with expressive individualism. Its anti-authority posture can easily become authoritarian. Someone must decide limits on self-determination, but who? And by what moral authority? Moreover, if desire is the driving force of expressivist ideology, can we trust desire to be as natural as we think? Is desire truly intuitive, or is it imitative? Further, are we aware of how much our desires are shaped and directed by algorithms designed to manipulate consumer behavior to boost profits? Dopamine, plain and simple, dictates much of our behavior, especially online.

Now is a moment ripe for Jesus's followers to expose the logical limits and self-contradictions of expressive individualism, to think more helpfully about desire and self and identity, and, ultimately,

---

13 Pamela Paul, "As Kids, They Thought They Were Trans. They No Longer Do," *New York Times*, February 2, 2024, https://www.nytimes.com/.

14 On this topic, see Rebecca McLaughlin, *The Secular Creed: Engaging Five Contemporary Claims* (Austin, TX: The Gospel Coalition, 2021), chap. 5.

to return to the individual and communal practices that counter Taylor's "individuating revolution." Now is the time for the church to build a countercultural vision of power and authority: as blessing, not curse.

The first step, of course, is to see that expressive individualism isn't "out there" but sitting in the pew. We need only notice how many Christians decide their lives today not by external forms of authority (the Bible, church leaders, generational and communal wisdom) but because of their "deeply felt personal insight."[15] According to Taylor, this translates as radical *religious* individualism. "The religious life or practice that I become part of must not only be my choice, but it must speak to me, it must make sense in terms of my spiritual development as I understand it."[16] In this spiritually expressivist house of cards, our emotions and intuitions confirm truth. If we participate in community at all (and many do not), our loyalties are often self-selecting along demographic, partisan, and ethical lines. We seek less to be transformed and more to be ratified. This is the tribalist pattern of our contemporary world, and its sorting is evident in church. Leaders then must be prodded to examine what unifies their congregations. Is it Jesus—or the political issue du jour?

As Christians, patterned after another way, truth, and life, we're called to radical resistance to this world, as Paul outlines in Romans 12:2. "Don't be conformed to the expressivist ideologies of this world," he might have said. "Be transformed by having your minds renewed with a growing capacity for greater discernment, that you might test what is good and acceptable and perfect."

Vice, of course, isn't something simply to defeat but to supplant. Sweep the house clean, Jesus said—but never leave it empty.

---

15 Taylor, *Secular Age*, 489.
16 Taylor, *Secular Age*, 486.

Resistance, as an act of repentance, will require both "putting off" sin (and the old self) and "putting on" virtue (and the new self in Christ). This requires leaders to speak a language foreign to their hearers, to say as gently and as unapologetically as possible, "Your intuitions can't always be trusted."

But to be clear, though sin mars and corrupts, it is not the *self* that is exactly the problem. It's not even *desire*, fashioned as we are in the image of a desiring God. No, for the *self* that has been made by God and is being remade in Christ, there is a renovation project underway. God is at work in us to will and to work for his good pleasure (Phil. 2:13). In the new heavens and the new earth, we will be recognizable, embodied "selves," and we will even bear marks of the cultural identities we carried with us on earth (Rev. 7:9). *Self*—as our received identity made by God and baptized in Christ and raised to walk in newness of life—is beautiful and good, and we can't grasp the heights of what God has planned for the restoration of all things, including our *selves*. When we see Jesus, those *selves* will bear his image as well as their own uniqueness.

Not Self-Made but God-Given

What needs to be "put off" in our age of expression is the myth of self-determination standing at the center of expressive individualism. According to the Bible, a "self-made" life is a fallacy, a farce—even in extreme cases like "famous for being famous" mega-influencer Kim Kardashian. Where Kardashian has been quick to credit hard work for her success, she has just as quickly been criticized for ignoring the wealth of privileges that made such success possible: Beauty. Wealth. Social connections.

As Christians, we must recover a humble acknowledgment of the divine "givenness" of our lives. It is God who quickens our breath, making possible every motion of being. As Paul tells his gathered audience in Acts 17, God ordains our lives in particular and material ways. This God has determined the "periods" of our lives and the "boundaries" of our dwelling places (17:26). Our moment in history, our birthplaces, our families of origin, our bodies: none of it is accidental, none of it self-determined. These "received" dimensions of our lives are means of grace by which people might "feel their way toward [God] and find him" (17:27). If we renounce the God-givenness of our lives, it's to our peril, not simply because the myth of self-making inflates our sense of grandiosity but because it prevents us from receiving God's good and wise providence.

Imagine if, during "testimony time" at church, everyone started even further back than a woman from my small group: "My grandmother was born in India . . ." (I was tired that night and confess I may have dozed a little as she spoke.) I see her wisdom now as I reflect on the way she chose to shape her narrative. God works generationally. I am not self-made, and you are not self-made. To remind ourselves of this, it might be good for us to start each day on our knees with a rehearsal of Psalm 100:3:

Know that the LORD, he is God!
It is he who made us, and we are his.

As we put off illusions of self-making, we put on worship and gratitude and praise.

These practices, which lift us beyond our blinding self-preoccupation, must be modeled for us in church, as God's people are called into his presence and reminded that "every good gift and every perfect gift is from above" (James 1:17). As congregations, we should regularly name what we've been given: in our buildings and neighborhoods, in our collective gifts and opportunities, even in our shared griefs and losses. We can remember that our corporate life and ministry are not the product of capital campaigns and talented preachers but *gifts*, "coming down from the Father of lights."

What's more, if we put off self-making, we must also put off self-sustaining. Often neglected in the critique of "expressive individualism" is the *individualism*. Postman is right to center his own history on America, land of the free. We are a people who conceive of freedom *from* tyranny and *for* self-rule. Enshrined in our national project is a bill of individual rights, freedoms the government promises to protect. These are immense democratic privileges, and I am grateful for them.

Still, the independent spirit of America can run counter to the communal vision of Christianity. We can become so taken with our own pursuit of happiness that we forget the practical keeping of our brother, our sister. Expressive individualism survives in isolation—and withers in healthy communities. There is no credible version of Christianity apart from Christ's church, and we're called out of the world and into local embodied community. Participation in a worshiping community counters the isolating effects of social media and its faux authenticity. To show up in person is to forgo carefully curated and performed versions of our lives; it is to foreground the collective adoration of Christ over the solitary pursuit of "likes" and affirming clicks.

At every opportunity, church leaders should promote the great good of embodied church life, both for ourselves and for the populations we serve: small groups and Sunday school, casseroles and corporate worship, food pantries and ESL classes. In his powerful sermon "The Expulsive Power of a New Affection," nineteenth-century Scottish preacher Thomas Chalmers argued persuasively that one way to fight sin isn't simply to tamp it down, as if smothering a fire.[17] We must rather kindle the flame of right affections. Viewed from this angle, a life-giving experience of a healthy church community would supplant our desire for thin, fragile online connections.

But this call to church participation is difficult to issue in an era when so much trust in the church has been lost. There has been documented abuse and systematic cover-up. There is all manner of political idolatry and abuse of power. It's hard to call people back to the church, especially to submit to its authority, when congregants have suffered at the hands of its leaders or suspect those leaders' true intentions.

Abuse of authority is real, and church leaders must promote cultures of transparency and accountability. But they must also preach and model why authority in the kingdom of God is meant as a good. This correction—of our culture's wholesale suspicion of authority—is an antidote to unbridled self-expression. Because I can't be trusted to *intuit* God's eternal good, I must therefore be *instructed*: by God's word and also by men and women of approved character, invested with an authority to teach, rebuke, admonish, and encourage. In many cases, these will issue blinkered warnings when my desires are wrong or wildly off base, or when they threaten self-destruction. This is the move of our gentle, loving Christ who, when the crowds were helpless and confused, taught them.

Today's problems aren't a call to give up on church, or even to give up on the God-given good that is cruciform power. There is Nietzsche's will to power—and there is the good shepherd,

---

17 Thomas Chalmers, *The Expulsive Power of a New Affection*, Crossway Short Classics (Wheaton, IL: Crossway, 2020).

laying down his life for the sheep (John 10:15). We need leaders formed in the servant love of Christ, who though he enjoyed all the prerogatives of divine power, took the form of a servant and "humbled himself by becoming obedient to the point of death, even death on a cross" (Phil. 2:8). Paradoxically, in God's kingdom, the way up is down. Power, in the hands of the world, is a sword. In the hands of Jesus, it's a plowshare, cultivating goodness and righteousness.

But, as Jesus asked, how might we know an act of true righteousness when we see one? For one, we aren't likely to see it on social media. As Jesus advised in the Sermon on the Mount and might reiterate in our world of performativity: *Do it in secret. Do it for the least. Persevere in small and ordinary faithfulness for the glory of God, and, if there is ever applause, credit the giver of every good gift who makes our acts of righteousness possible.*

Made in the image of God, humans are destined for glory. But this glory, outside of Eden, has fallen into disrepair, and no achievement or celebrity can redeem the "self" we have lost. Perhaps we're all now looking to recover it. We try to make a self—and lose a soul along the way.

It is tempting to mute the human voice, given the wrong notes it tends to sing. But this would be an overcorrection, even a capitulation in the inhuman age of AI. Jesus is calling to all of us who are wrung out by the wearying efforts of self-making and the craven demands of our public audiences: "Come to me." According to the gospel, in Christ, self-expression can be redeemed and the voices of humans lifted up—in chorus with all creation—to declare not their own glory but God's.

With the rocks, together we can learn to cry out, now and into eternity (Luke 19:40).

Discussion Questions

1. How has the potent longing for "expressive individualism" informed the way our smartphones are set up? And, on the flip side, how has the way our smartphones are set up informed our potent longing for "expressive individualism"?

2. An emphasis on a particular value is often a pushback against another value. What do you think our present prioritization of self-expression is pushing back against? Are there any positive elements a culture of self-expression has given us?

3. Michel contends that "life itself is 'performed' on social media today." Discuss the tension between the pressure to "be authentic" and the pressure to "perform" for the online community. Have you felt these conflicting pressures before? Can users avoid this tension when posting content for large audiences online—or is it unavoidable?

4. What theological concepts, scriptures, or names of God might the church need to emphasize in response to our current proclivity toward being "self-made"? How can Christians argue it is better that we are *not* self-made?

4

The Origins and Implications
of a Post-Truth World

Hans Madueme

"WE'RE ALL MAD HERE. I'm mad. You're mad."[1] The Cheshire
Cat utters these words to Alice in Lewis Carroll's *Alice's Adventures
in Wonderland*, but he might as well be talking to us.

Indeed, the world has gone mad. A growing contingent of people
espouse a flat earth; millions of transgender ideologues sincerely
believe men can become women and women can become men;
countless people are duped by digital deceptions and unknow-
ingly spread misinformation. To complicate matters, "conspiracy
theories" sometimes contain kernels of truth or later prove not to
be wildly off the mark. Half-truths, untruths, and "true enough"
statements bombard us every day in the digital age. But trustworthy truth, verifiable truth, and solid truth? Those are hard to find.

1 Lewis Carroll, *Alice's Adventures in Wonderland and Through the Looking-Glass* (1871; repr.,
New York: Bantam Dell, 2006), 50.

How did we get here?

Some argue the internet democratized information and authority, killing the "expert" class and rendering expertise meaningless. The COVID-19 pandemic and growing media polarization have accelerated this trend.[2] Others emphasize American Christianity's anti-intellectualism and distrust of secular science to explain the complicity of many evangelicals in "post-truth" discourse.[3]

Such responses are all insightful, but Neil Postman had already prophesied what was coming. Decades ago, he warned that television was distracting us with trivia, infantilizing public discourse, and reducing our national life to a circus. In that milieu, he wrote, "culture-death is a clear possibility."[4] Alas, we did not take him seriously—or not seriously enough. The chickens have now come home to roost.

Postman's ideas can help us understand three reasons we as a society (and church) have become discombobulated in a post-truth age: (1) the shift from modernism to postmodernism, (2) the shift from television to social media, and (3) the shift from public democratic safeguards to private financial corporations.

On Truth

At the outset, let's be clear on what we mean by truth.[5] The Lord is the true God (Jer. 10:10), and his eternal Son embodies that truth

---

2   Tom Nichols, *The Death of Expertise: The Campaign against Established Knowledge and Why It Matters*, 2nd ed. (New York: Oxford University Press, 2024).

3   Josh A. Reeves, *Redeeming Expertise: Scientific Trust and the Future of the Church* (Waco, TX: Baylor University Press, 2021). See also Nathan O. Hatch, *The Democratization of American Christianity* (New Haven, CT: Yale University Press, 1989).

4   Neil Postman, *Amusing Ourselves to Death*, 20th anniversary ed. (1985; repr., New York: Penguin Books, 2005), 156.

5   For much of what follows in this section, I'm indebted to Arthur Holmes, *All Truth Is God's Truth* (Downers Grove, IL: InterVarsity, 1977).

(John 14:6). Truth inheres in the triune God and his own intrinsic perfection. Since God created all things, his wisdom is the source of truth about everything. Although believers see through a glass darkly, the truth itself is unified in the wisdom of God.

Unbelievers may deny God intellectually, but they cannot deny his creation. God made them in his image and gave them an intelligible world to inhabit. They too can perceive truth imperfectly. They can know that two squared is four, that two contradictory claims cannot both be true, that Argentina beat France in the 2022 World Cup. All of us can know proximate, contingent truths (facts!) through deduction and induction. Regardless of where we find it, all truth is God's truth—and this truth is both absolute and personal.

Truth is absolute for it is unchanging and universal. As Arthur Holmes remarks, "If God's knowledge is complete and perfectly true, then truth itself cannot change; it remains the same for every time and place in creation; it is absolute."[6] Yes, the weather and presidents are always in flux, but even contingent truths are unchanging for any specific place and time. The weather right now in my neighborhood is sunny, which is true irrespective of the weather tomorrow.

Truth is personal, not merely propositional. Scripture's propositions are true because God is utterly trustworthy. We believe Scripture because we trust God—"Your word is a lamp to my feet and a light to my path" (Ps. 119:105). The biblical view of truth unites the abstract and the personal. God calls us to do the truth: "If we say we have fellowship with him while we walk in darkness, we lie and do not practice the truth" (1 John 1:6).

6   Holmes, *All Truth*, 32–33.

Jesus lived that way and holds all the treasures of wisdom and knowledge (Col. 2:3). All truth finds its unity and coherence in the person of Christ.

Shift 1: From Modern Rationalism to Postmodern Backlash

The rise of Enlightenment rationalism and modern science in the seventeenth century broke down this unified notion of truth. Truth became reduced to objectivity without the personal side—hence works like Voltaire's satirical novella *Candide* (1759), David Hume's *An Enquiry Concerning Human Understanding* (1748), and Immanuel Kant's essay "What Is Enlightenment?" (1784). Empirical science was the new standard of knowledge, with little room for personal faith in a divine person as grounding for truth.

In the nineteenth century, unsurprisingly, the Romantic movement reacted against this imbalanced emphasis on scientific objectivity and redirected focus to the emotions in art, music, literature, and the like. Thus, since the Enlightenment, the story of truth in the West moved back and forth "between the demand for logical proof or objectively verifiable statements of fact, on the one hand, and the romanticist and existentialist reactions, on the other."[7] Amazingly, this shaky view of truth still held society together into the twentieth century. But the foundations of Enlightenment rationalism and modernist trust in science were starting to crumble.

Friedrich Nietzsche blew the whistle early on. He realized Enlightenment thinkers were trying to have their cake and eat it too. They banished God from the modern world while trying to preserve principles like truth, rationality, and order. But Nietzsche saw that if God is dead, then society's old pillars were equally doomed. In

---

7   Holmes, *All Truth*, 7.

Nietzsche's parable "The Madman," the atheist says, "Who gave us the sponge to wipe away the entire horizon? What did we do when we unchained this earth from its sun? Whither is it moving now? Whither are we moving now? Away from all suns? Are we not plunging continually? Backward, sideward, forward, in all directions? Is there any up or down left?"[8]

Nietzsche foresaw twentieth-century postmodernism and the ideas of Michel Foucault, Jean-François Lyotard, Richard Rorty, Jean Baudrillard, and many others. These postmodernists were all saying we can no longer believe in absolute truths; there is no reality prior to interpretation. Meaning always depends on context and subjective perspective. Science, ethics, morality, politics—indeed, everything—is tainted by human subjectivity. The epistemology of modernism implies "reality pre-exists to be discovered," whereas for postmodernism we construct reality "through subjective discourse and interpretation."[9] Postmodernism paved the way for our post-truth era.

Shift 2: From Television to Social Media

In 2016, Oxford Dictionaries named *post-truth* the word of the year, which they defined as "relating to or denoting circumstances in which objective facts are less influential in shaping public opinion than appeals to emotion and personal belief."[10] According to Lee McIntyre in his book *Post-Truth*, "It is an expression of concern by

8   Friedrich Nietzsche, *The Portable Nietzsche*, trans. Walter Kaufmann (New York: Viking, 1968), 95.

9   Susana Salgado, "Online Media Impact on Politics: Views on Post-Truth Politics and Post-postmodernism," *International Journal of Media and Cultural Politics* 14, no. 3 (2018): 321, https://repositorio.ul.pt/bitstream/10451/35569/1/Salgado_Susana_Posttruth_IJMCP.pdf.

10 "Word of the Year 2016," Oxford Languages, accessed June 26, 2024, https://languages.oup.com/.

those who care about the concept of truth and feel that it is under attack." In a post-truth world, truth is vulnerable because it is seen as irrelevant.[11] Journalists can't be trusted—it's "fake news." Climate science and vaccines are bunk—it's fake science. The Holocaust never happened—it's fake history. And Russian trolls create fake Americans on social media, and bots generate fake followers and social media "likes."[12] Welcome to the post-truth hellscape where everyone and everything is suspect.

Media Shapes Our Lives

Postman's *Amusing Ourselves to Death* sounded the alarm decades ago, insisting that television transforms the nature of public discourse. How else do we explain the high number of celebrities gaining political office in recent history, not just Ronald Reagan but Jesse Ventura, Arnold Schwarzenegger, Al Franken, Donald Trump, and others?

Postman was an early proponent of *media ecology*, which "tries to find out what roles media force us to play, how media structure what we are seeing or thinking, why media make us feel and act as we do."[13] The book medium invites readers into analytic, reflective,

---

11 See Lee McIntyre, *Post-Truth*, MIT Press Essential Knowledge (Cambridge, MA: MIT Press, 2018), 6.

12 Michiko Kakutani, *The Death of Truth: Notes on Falsehood in the Age of Trump* (New York: Penguin Random House, 2019), 13. Post-truth discourse is increasingly driven by authoritarian governments manipulating what people see and read on social media. For example, see Peter Pomerantsev, *Nothing Is True but Everything Is Possible: The Surreal Heart of the New Russia* (New York: Public Affairs, 2014); and Louisa Kim, *The People's Republic of Amnesia: Tiananmen Revisited* (New York: Oxford University Press, 2014).

13 Carlos A. Scolari, "Media Ecology: Exploring the Metaphor to Expand the Theory," *Communication Theory* 22, no. 2 (2012): 205. Over fifty years ago, Postman wrote, "[Media ecology] looks into the matter of how media of communication affect human perception, understanding, feeling and value; and how our interaction with media facilitates or impedes our chances of survival." Neil Postman, "The Reformed English Curriculum," in *High School*

critical reasoning: "To engage the written word means to follow a line of thought," Postman wrote, "which requires considerable powers of classifying, inference-making and reasoning." The media ecology of reading, he continued, tends to "uncover lies, confusions, and overgeneralizations, to detect abuses of logic and common sense."[14]

Not so with television. It operates by the logic of entertainment, Postman argued. Television's currency is visual delight rather than thoughtful action, engaging images rather than textual argument. Postman wasn't a prude inveighing against entertaining television; rather, he objected to reinventing public discourse in television's image, where news and politics become trivialized as just another form of entertainment.

Consider television debates. They are all about the thrill, excitement, and spectacle, not at all about critical thought or serious dialogue. Postman argued that "embedded in the surrealistic frame of a television news show is a theory of anticommunication, featuring a type of discourse that abandons logic, reason, sequence and rules of contradiction."[15] In Postman's view, television in the 1980s was already leading us to lose "our sense of what it means to be well informed."[16] It was already sowing the seeds of our present post-truth world.

1980: The Shape of the Future in American Secondary Education, ed. Alvin C. Eurich (New York: Pitman, 1970), 161.

14 Postman, *Amusing Ourselves*, 51.

15 Postman, *Amusing Ourselves*, 105. cf. C. John Sommerville, *How the News Makes Us Dumb: The Death of Wisdom in an Information Society* (Downers Grove, IL: InterVarsity, 1999).

16 Postman, *Amusing Ourselves*, 107–8. Postman also lamented the modern tendency to equate information with real knowledge: "This is the elevation of information to a metaphysical status: information as both the means and end of human creativity. In Technopoly, we are driven to fill our lives with the quest to 'access' information. For what purpose or with what limitations, it is not for us to ask; and we are not accustomed to asking, since the problem

Social Media Raises the Stakes Even Higher

Postman wrote well before the advent of Facebook and X, but his warnings are even more relevant today. As Hossein Derakhshan states, "Social media represents the ultimate ascendance of television over other media. . . . [It] not only shares many of TV's ills but also creates new ones."[17] While television can still surprise and challenge its viewers, algorithmically driven social media tends to affirm and reinforce what we already believe. X's bite-sized communication format—especially the original 140- and then 280-characters limit—assumes debate is always simple and easy. The underlying epistemology of tweets presumes solutions to social, political, or religious problems are brief, superficial, disembodied, and entertainment-driven, favoring monologue over dialogue and ad hominem over thoughtful reflection.

Social media's ascendancy has accelerated the decline of traditional news media (which was already struggling in Postman's day). American newsroom employment has dropped precipitously over the past two decades, creating "news deserts" where local news has vanished and communities depend on social media and national reporting (e.g., the *New York Times* and the *Wall Street Journal*).[18]

While social media is ill-suited for nuanced reporting and fostering meaningful discourse about substantive issues, it excels at fanning the flames of partisanship and exacerbating the "us vs. them" narratives that now dominate politics. Social media rewards

---

is unprecedented. The world has never before been confronted with information glut and has hardly had time to reflect on its consequences." Postman, *Technopoly: The Surrender of Culture to Technology* (New York: Vintage Books, 1992), 61.

17 Hossein Derakhshan, "Social Media Is Killing Discourse Because It's Too Much Like TV," *MIT Technology Review*, November 29, 2016, https://www.technologyreview.com/.

18 Michael Bugeja, "Fact to Fake: The Media World As It Was and Is Today," in *Media, Technology and Education in a Post-Truth Society*, ed. Alex Greach (Bingley, UK: Emerald, 2021), 118.

incendiary statements and discourages careful rhetoric, transforming politics into a schoolyard where bullies who make the loudest noise attract large followings. No wonder "trolling" has become common practice for leaders and politicians at the highest levels.[19]

In a post-truth epistemology, popularity is sexier than truth: "Lengthy, detailed disquisitions do not fare very well against short, biting sarcasm. They also do not fare well against comments that, however inane, rack up a far greater number of likes."[20] Social media creators want us to keep coming back for more. "The outcome is a proliferation of emotions," cautions Derakhshan, "a radicalization of those emotions, and a fragmented society. This is way more dangerous for the idea of democracy founded on the notion of informed participation."[21]

All in all, the transition from television to social media has had a devastating effect on the information's trustworthiness and the possibility of shared truth. In the social media world, algorithms feed audiences what they want, not what they need. Politicians and journalists increasingly speak to get clicks and go viral, not to advance truth. And the onset of AI technology, which can put any words in anyone's mouth in remarkably real-looking video or audio clips, gives us new reason to doubt online content's trustworthiness.

Shift 3: From Public to Corporate Interest

The apostle Paul said, "The love of money is a root of all kinds of evils" (1 Tim. 6:10). This lust for mammon is a key element in the

---

19 For insightful analysis, see Jason Hannan, "Trolling Ourselves to Death? Social Media and Post-Truth Politics," *European Journal of Communication* 33, no. 2 (2018): 214–26.

20 Hannan, "Trolling Ourselves," 220.

21 Derakhshan, "Social Media is Killing Discourse."

post-truth crisis. Here, too, Postman saw the writing on the wall: "The Founding Fathers did not foresee that tyranny by government might be superseded by another sort of problem altogether, namely, the corporate state, which through television now controls the flow of public discourse in America."[22]

Postman recognized that corporations profit from consumers amusing themselves to death. However, Postman's critics often dismissed him as a bloviating, melodramatic Luddite unwilling to see anything good in modern technology. This broadside is dead wrong: ironically, the money trail suggests Postman should have laid it on thicker.[23]

Private Corporations Triumphed over Public Interest

Here we must review recent history. While the First Amendment to the US Constitution protects the freedom of the press, it does not compel journalists to protect democracy. In the early twentieth century, the news industry was dominated by large media conglomerates shaping political discourse and stifling competition.

During Franklin Roosevelt's presidential administration, successful lobbying by concerned citizens led Congress to pass the Communications Act of 1934. This legislation created the Federal Communications Commission (FCC), which regulated radio, limited how many stations businesses could own, and implemented other measures in service of the public interest.[24] Then in 1949,

---

22 Postman, *Amusing Ourselves*, 139.

23 My remarks in this section closely follow Nolan Higdon and Mickey Huff, *United States of Distraction: Media Manipulation in Post-Truth America (And What We Can Do About It)* (San Francisco: City Lights Publishers, 2019), 37–52.

24 See United States, "Communications Act of 1934," FCC website, https://www.fcc.gov /Reports/1934new.pdf.

protests after World War II prompted Congress to pass the fairness doctrine, ensuring broadcasters gave equal airtime to different perspectives. These initiatives were driven by the belief that democracy thrives when we privilege public over private interests, when citizens are exposed to various perspectives rather than one monolithic view dominating the media.

Unsurprisingly, the private sector retaliated against these public-interest measures. James M. Buchanan (1919–2013), a former economics professor, perceived the government and labor unions as a threat against individual liberties. Backed by corporate donors, he "began a trend of constructing brain trusts on college campuses to raise money, hire conservative faculty, and produce pro-business content that would shift educational emphasis from liberal notions of the public good to more right-leaning views that regard government as an obstacle to personal freedoms."[25] Through trial and error, he managed to reframe the issue as a debate between governmental overreach (bad) versus free choice (good).

Buchanan's efforts to promote capitalism and strengthen private interests inspired the 1971 Powell Memo, which opposed government regulation of media and academia.[26] Ronald Reagan defended the same corporate principles during his presidency. These corporate efforts eventually convinced Americans that governmental intervention impedes economic growth and prosperity for everyone: "The business-dominant economy that gradually emerged invested heavily in the creation of a veritable one-party

---

25 Higdon and Huff, *United States of Distraction*, 41. For additional documentation, see Nancy MacLean, *Democracy in Chains: The Deep History of the Radical Right's Stealth Plan for America* (New York: Penguin, 2017).

26 The author of the memo was Lewis J. Powell, a lawyer at the time, who would later be appointed to the US Supreme Court.

political system—the pro-corporate party—with two factions, the Republicans and Democrats, funded to uphold corporate interests above all others."[27]

The Post-Truth Jungle Feeds on Deregulation

There is a direct link between corporate interests and our post-truth culture. Once the two main US political parties had been co-opted by the corporate class, news media organizations were systematically deregulated. One federal policy after another fell as corporations lobbied relentlessly to axe as many of them as possible.

The Telecommunications Act of 1996 reversed the Communications Act of 1934; as a result, "the fifty corporations that owned the bulk of the media during the 1980s [were] reduced to six by 2012: News Corp, Disney, Viacom, Time Warner, CBS, and Comcast."[28] By 1987, Congress abolished the fairness doctrine under pressure from corporate lobbyists, opening the way for commercial interests to supersede traditional journalistic reporting. As federal regulation fell, the wealth of the remaining media corporations skyrocketed.

Post-truth culture is financially lucrative for media corporations and pundits across the political spectrum who capitalize on polarizing rhetoric. Both liberal and conservative media outlets have benefited from controlling media narratives to fragment audiences and create angry tribes eager to click on more, watch more, and read more, all to reinforce their fears and antagonism toward "the other side." When the news industry is incentivized primarily by corporate profits, and little by what's in the public interest, we shouldn't be surprised that communicating truth becomes less important than storytelling: framing what's newsworthy, and how

27 Higdon and Huff, *United States of Distraction*, 45.
28 Higdon and Huff, *United States of Distraction*, 47.

the news is reported, in terms of reinforcing established narratives. Accurate, nuanced reporting doesn't get clicks. Narrative-fueling "news" does.

In a landscape like this, truth disintegrates into political spectacle and rabid polarization.

Navigating Post-Truth Culture with Wisdom

The post-truth culture has all the hallmarks of the church's three-headed foe—the world, the flesh, and the devil. Post-truth discourse fits the pattern of this world as it draws Christians away from the Father's love (Rom. 12:2; 1 John 2:15–17). The mean-spirited, often vicious rhetoric that proliferates online gratifies the cravings of the flesh (Eph. 2:3) and breeds habits antithetical to the fruit of the Spirit in Galatians 5:22–23. Anyone who conforms to the post-truth script plays into the hands of the devil—a liar and the father of lies (John 8:44).

But what to do? The forces are so big, the issues so complex, that the problem feels like a giant killer octopus: impossible to evade, much less overpower. And we can expect things to worsen as generative AI floods the information ecosystem with convincingly fabricated content—spreading misinformation and undermining expertise (e.g., deepfakes)—that exacerbates the epistemological crisis and further erodes trust in traditional sources of truth.

And who knows what will come after the post-truth era? Maybe the chaos, grief, and loneliness will lead people to recognize the need for transcendent sources of truth. Or maybe the post-truth world will become so normalized that the label will grow antiquated, since "truth" will have become a relic of a bygone era.

Even if we don't know where contemporary culture will go from here, do not lose hope. Christians follow Jesus, who embodies truth and calls us to be people of the truth. As Paul says, "Having put away falsehood, let each one of you speak the truth with his neighbor, for we are members one of another" (Eph. 4:25). Lean into this gospel reality so you always cling to the truth, testify to it daily, and shun the post-truth theatrics around us.

God gave us Scripture to keep us rooted in timeless truth. Rather than taking the Old and New Testaments for granted, cultivate habits of biblical literacy, feeding regularly on the words of God (Matt. 4:4). The Bible is the ultimate test of truth and check against falsehood. The more steeped we are in God's wisdom revealed in Scripture, the more discerning we'll be as we navigate a post-truth world.

Yet we are whole people, not brains on sticks. Our actions and choices are not solely based on careful, solitary deliberation. The post-truth crisis has taught us that Enlightenment rationalism doesn't fully capture our nature as embodied, relational creatures: "Belonging is stronger than facts."[29] People still live falsely even while "knowing" the truth—because we are powerfully catechized by the company we keep (1 Cor. 15:33). Thus we need healthy communities that will nurture us to treasure what is true, including people from church, close friends, neighbors, and family.

As a key part of your formative community, keep company with trusted teachers from the past, reading old books, leaning on ancient Christian voices to give us perspective on the untested present. Build a reading diet that isn't imbalanced toward the "new" but includes healthy doses of the "old." Cherish perspectives that

---

29 Zeynep Tufekci, "How Social Media Took Us from Tahrir Square to Donald Trump," *MIT Technology Review*, August 14, 2018, https://www.technologyreview.com/, cited in Jeffrey Bilbro, *Reading the Times: A Literary and Theological Inquiry into the News* (Downers Grove, IL: IVP Academic), 151.

have been tested by time and have endured as resonant insights over decades, centuries, and millennia.

Finally, slow down. When processing and sharing information, speed is often the enemy of truth ("time-tested" is a phrase for a reason). Truth vetted across time is generally more reliable. Part of our struggle with truth in the social media era is the high premium on speed; information is shared and goes viral before its veracity is fully tested, before the facts are known and understood. Even trustworthy journalists and experts can stumble here. But hear what Proverbs 2:3–5 says about the pursuit of wisdom:

> Indeed, if you call out for insight
>   and cry aloud for understanding,
> and if you look for it as for silver
>   and search for it as for hidden treasure,
> then you will understand the fear of the LORD
>   and find the knowledge of God. (NIV)

According to Proverbs, having wisdom is thinking and living in line with how things actually are—the truth—which takes time and work. Christians can mitigate the epistemic effects of a sped-up world by intentionally slowing down, asking questions, investigating sources, and doing research rather than taking claims at face value.[30] And many of us need to step away from particular social media platforms, for a season or even for good. It may sound radical, but this could be the remedy we need to stay true to Christ in a world gone mad.

---

30 For helpful resources, see Cal Newport, "It's Time to Dismantle the Technopoly," *New Yorker*, December 18, 2023, https://www.newyorker.com/; Cal Newport, *Digital Minimalism: Choosing a Focused Life in a Noisy World* (New York: Penguin, 2019); Jason Thacker, *Following Jesus in a Digital Age* (Brentwood, TN: B&H, 2022); Bilbro, *Reading the Times*, esp. 143–74.

Discussion Questions

1. Of the three explanations Madueme offers for how we got to a post-truth world, which do you see as the most significant? Are there other factors you believe have led to this point?

2. Where have Christians been complicit in the dynamics of an online world increasingly full of lies, misinformation, conspiracy theories, and partisan narratives? How can we model more careful, discerning behavior online?

3. How does the fragmented, algorithmic structure of online media play a role in society's growing loss of consensus and shared truth? Have you had an experience where you realized the "facts" you were being fed on your feed were different from those others in your life were receiving?

4. What habit or behavioral tweak can you implement now that will better equip you to discern fact from fiction and truth from falsehood?

5

Striving for Seasonableness
in a "Now . . . This" World

Samuel D. James

IN *CATCHING FIRE*, the second film in the Hunger Games trilogy, head gamemaker Plutarch Heavensebee (Philip Seymour Hoffman) explains, in a brief sequence, the emotional power of context. Ruminating on how best to destroy the story's heroine, Katniss Everdeen (Jennifer Lawrence), Heavensbee suggests that President Snow (Donald Sutherland) could turn the people against her by pairing together footage of Katniss's luxurious wedding with scenes of oppression and violence: "She's engaged. Make everything about that. What kind of dress will she wear? Floggings. What's the cake going to look like? Executions. Who's going to be there? Fear. . . . Show them she's one of us now."[1]

---

1   *The Hunger Games: Catching Fire*, directed by Francis Lawrence (Lionsgate Films, 2013).

Katniss is a warrior and populist hero who has defied the tyrannical government. Her victories inspire solidarity throughout the land. There's nothing her enemies can say or argue that will dampen her support. But Heavensbee understands that it's not always necessary to prove something is true—or even to say something is true. Sometimes, all you have to do is blur the lines between ideas enough that people will be unable to tell where one ends and the other begins.

Media is powerful because it directs our attention. For the television viewer, there is no reality other than what is on-screen— the camera shapes the audience's sense of reality. In his *Amusing Ourselves* chapter "Now . . . This," Neil Postman observed that television's inevitable blending of radically disparate content communicated a message about the possibility of coherent thought. The expression "Now . . . this," particularly in the context of news coverage, is "a means of acknowledging the fact that the world as mapped by the speeded-up electronic media has no order or meaning and is not to be taken seriously. There is no murder so brutal, no earthquake so devastating . . . that it cannot be erased from our minds by a newscaster saying, 'Now . . . this.'"[2]

Postman argued that the advent of mass media—communication and journalism that transcended geographic boundaries and created the sense of a "national" news-consciousness—promised an inevitable slide into superficial discourse. It wasn't enough for people in a small town to have media institutions covering information relevant to them. Instead, the possibility of accessing news from faraway places and utterly remote individuals put value on information *for its own sake*. But because the small-town viewer cannot do anything with this far-flung news, the only choice is to

2 Neil Postman, *Amusing Ourselves to Death*, 20th anniversary ed. (1985; repr., New York: Penguin Books, 2005), 99.

consume quickly and move on. For us to do this efficiently, the news must be easily consumable, and then easily disposable—making way for something else, something next. "Now . . . this."

This sort of rapid, superficial succession of information, designed to be discarded, is not just a neutral presentation choice. It is a worldview itself. The vast majority of its consumers will be unable to think clearly or respond appropriately. It is, according to Postman, "a world of fragments" whose ethic is "not coherence but discontinuity."[3] This media habitat resists efforts to apply logic and reason. We are instead left with knee-jerk reactions ("That's outrageous!") and a hazy sense of what kind of world we live in.

Could there be any better description of the internet age?

Everything, All of the Time

In 2021, as part of his pandemic-themed album *Inside*, comedian and filmmaker Bo Burnham released a song titled "Welcome to the Internet." The track—a surreal homily on the dizzying diversity of the web—highlights the way the internet's whimsical content and its violence and paranoia are all just a click away from one another. The cumulative effect of our experience online amounts to "a little bit of everything, all of the time."[4]

It would be difficult to find a better anthem for the internet. Before it is anything else, the web is an endless ocean of *stuff*. There are no meaningful limits on what an online device can summon. Any question, any search, any diversion, any kink, any craving—no matter how niche, how peculiar, or even how destructive—is available in the world of the internet.

---

3 Postman, *Amusing Ourselves*, 110.
4 Bo Burnham, "Welcome to the Internet," February 12, 2022, YouTube video, https://www.youtube.com/watch?v=k1BneeJTDcU.

The internet as a technology doesn't recognize any spectrum of wholesomeness or usefulness. A video showing how to replumb a kitchen sink is every bit as accessible as a video showing violent pornography. Articles that contemplate the most challenging questions of life, God, and society show up on a search results page alongside advertisements for shapewear or Fazoli's. Algorithms—the controlling power of much of the modern internet—possess no human dilemmas like a finite amount of time or a need for virtue. They recognize no meaningful difference between a template for writing a cover letter or step-by-step instructions for building a bomb.

Not only is the web an endless ocean, but it is in many ways a borderless ocean. Whereas a physical place like a school or a church exists as a single environment that offers a select list of experiences and ideas, the web has no natural boundaries. I may be at a certain website for a particular reason, but an advertisement, a hyperlink, or an embedded video can instantly transport me to another, radically different place. This effect is magnified by social media, the way millions of people now navigate the web's sprawl. A fifteen-minute scrolling session can yield everything from inspirational quotes to racist tirades to advertisements to news of a loved one's cancer diagnosis.

The internet's borderless width and bottomless depth takes the "Now . . . this" effect of television to an extreme. Our response to it is both intellectual and emotional.

Intellectually, the web's infinite porousness makes it difficult to think clearly. In *The Shallows: What the Internet Is Doing to Our Brains*, Nicholas Carr offers convincing evidence that the human brain's capacity to comprehend language is compromised in a computer medium.[5] Digital technology's ethic of speed, efficiency,

5 Nicholas Carr, *The Shallows: What the Internet Is Doing to Our Brains* (New York: Norton, 2010).

and varied input shapes how information is presented online and how we interpret it. Hypertext, advertisements, notifications, and links all condition how we understand the ideas presented to us on the web.

Imagine the following scenario: You are watching a YouTube creator's video about the health dangers of a particular food. In a five-minute span, the host combines personal experience ("I experienced these horrible symptoms while eating that food"), references to data ("1 in 3 people who consumed these foods developed a serious illness"), disturbing pictures of food factories, a list of politicians who receive donations from this corporation, and a link to the host's recommended book.

How would you begin to parse the claims of that video? You could, theoretically, check to see where the data came from, do research to see if those pictures were real, investigate the political donations, and read the book. But a mere five-minute video will almost never encourage such fact-finding. Instead, what is likely to happen is you will either commiserate with the host's personal experiences or not, experience revulsion at the photos or the names of politicians or not, and decide whether you like this particular content creator or not. From there, you may come away with a vague sense of agreement or disagreement.

The web's jumbled nature also has implications for us emotionally—affecting our ability to feel appropriately. Sociologist Jean Twenge and social psychologist Jonathan Haidt have written at length on social media's strong correlation with mental health problems, especially in teens.[6] While explanations are still quite theoretical at this point, one plausible factor is the internet's

---

6   See Jean Twenge, *iGen: Why Today's Super-Connected Kids Are Growing Up Less Rebellious, More Tolerant, Less Happy—and Completely Unprepared for Adulthood* (New York: Atria,

tendency toward performative behavior—posting, liking, sharing, all so others can see you did—cultivates anxiety and exhaustion.

The online world bundles thinking and belonging so tightly together that we are not sure where one ends and the other begins. Comments are presented right alongside content, so we often see what people are saying before we even see what they are saying it about.

Research shows that comments, especially extreme ones, can change how internet users perceive an argument.[7] In the flow of discourse online, arguments intertwine with reactions, expertise blurs with amateur opinion, and an outsized reaction to any given thing becomes a story in its own right (morning talk shows often devote daily segments to reporting on viral pseudo events as if they're breaking news). Information and reactions to information online are all of a piece, such that information's "value" starts to become tied to the size and enthusiasm of the reaction it sparks.

Because of these dynamics, social media corporations know that provocative content eliciting negative emotion (such as anger or depression) captures attention—and fuels response comments and back-and-forth arguments—better than other content. Naturally, it's privileged by algorithms.

Striving for Seasonableness

The numbing effects of endless digital content can feel overwhelming. Are we confined, as residents of an online age, to mental blurriness and emotional fatigue? I don't think so. There is little that

---

2017), and Jonathan Haidt, *The Anxious Generation: How the Great Rewiring of Childhood Is Causing an Epidemic of Mental Illness* (New York: Penguin, 2024).

7   Dominique Brossard and Dietram A. Scheufele, "This Story Stinks," *New York Times*, March 2, 2013, https://www.nytimes.com/.

ordinary Christians and churches can do about our media landscape (although we ought not discount the possibility of Christians in technological spaces working for real change). But there is much we can do to reclaim wisdom and peace as those made in God's image, bearing witness to Jesus in a hostile world.

Biblical Wisdom Literature teaches about the idea of "seasonableness"—the fittedness of things, the beauty and clarity that comes when things find their proper place. Seasonableness reflects the wisdom of God, who created the world and imposed a divine order on it so everything was in its proper place, doing its proper job.

When God rested on the seventh day of creation, he instituted the Sabbath, setting in motion the seasonableness of labor for six days and the seasonableness of rest on the seventh (Gen. 2:2–3). He created Eve as a helper "fit" for Adam (2:18). In Ecclesiastes 3, Solomon famously contemplates the seasonality of life in God's world: "For everything there is a season, and a time for every matter under heaven" (3:1). Everything in its proper place.

Elsewhere, Proverbs commends seasonableness in speech: "A word fitly spoken is like apples of gold in a setting of silver" (25:11). In the Song of Solomon, the reader is admonished not to "awaken love until it pleases" (8:4), while a sluggard is to consider the way the ant uses summer to its advantage so winter will not destroy it (Prov. 6:8). The alternative to seasonableness is foolish chaos: a failure to understand the need of the hour, as well as an impulsive blindness to long-term consequences.

Seasonableness offers a clue as to how we as Christians can resist the spiritual fog of the "Now . . . this" age. Digital culture's onslaught of content presents us with a chaotic, impulsive, haphazard sense of the world. Users are conditioned to accept that whatever the algorithm feeds them is perfectly appropriate. But this indiscriminate consumption often confuses and frustrates us. Seasonableness asks, "Is *this* medium the best way to think about *this* issue?" or "What deserves my attention and energy *right now*?"

Because it's rooted in the orderliness of God's design for his world, seasonableness results in clarity rather than confusion. One of the ways we can apply seasonableness to our digital habits is to identify particular formats that seem prone to undermining deep, reflective thinking.

Many have noticed that TikTok, for example, seems to be a platform of choice for those deconstructing their faith.[8] It's hard to ignore the fact that TikTok privileges short bursts of information and rewards attention-catching content that elicits strong emotions like anger. Seasonableness asks whether such a medium is appropriate for serious conversations about faith and doubt. Christians struggling to answer deep questions should consider how face-to-face conversations, thoughtful books, and other resources are more naturally conducive to slow, meaningful thinking.

Similarly, YouTube is a highly effective place to find DIY tips. But is it equally effective at offering practical life counsel? Typing in a search like "Should I get married" or "Is therapy good for me" will yield thousands of results, some better than others. But these are questions tied to both transcendent truths and particular life details. A YouTube influencer cannot answer these questions effectively for people they don't know (99 percent of their viewers). Seasonableness discerns what digital media does and doesn't do well.

Seasonableness is also concerned with the time and place of our digital habits. Constant connectivity, the default for most people with a smartphone, can hardly resist the "Now . . . this" effect.

---

8 Renee Roden, "TikTok Emerges as Venue to 'Deconstruct' Catholic Theology," *National Catholic Reporter*, February 14, 2022, https://www.ncronline.org/.

A morning devotion, hours of work, personal conversations, meals, and bedtime are all daily assaulted by news and notifications. This isn't only disruptive; it is deadening. The nonstop stretching of our attention span depletes us mentally and emotionally, keeping us in a state of anxiety.

The principle of seasonableness can help. It's the difference between a mindless scroll out of boredom or apathy and an intentional use of digital media to learn or share. Interrogating your use of digital media might yield surprisingly effective habits and good questions: Do I know why I'm logging on right now? Is it to find something, share something, or accomplish a task? Or am I merely seeking distraction? Embedded in these questions is the hope that our digital media intake can be purposeful rather than aimless, intentional rather than passive.

Social media is a frequent "unwinding" activity for many. Yet it's not actually that restful. Social media's passivity is different from leisure, which has a tendency to refresh us by helping us to stop focusing so much on ourselves and our own mental state and instead pay attention to something pleasant outside us. This partially explains why an hour of scrolling Instagram or Facebook feels oddly tiring. What keeps us locked in for that amount of time is often aimlessness. But aimlessness can be identified and resisted by asking, "What is the best use of my attention, my energy, and my affection *right now*?" Answers will vary, because life's seasons vary.

Seasonableness as a discipline awakens us to the reality of providence. God, by his kind and gracious authority, has declared a certain set of circumstances and roles for all of his people. Our flight into digital distraction is often Jonah-like; we want to forget these circumstances and delegate those roles. *Escapism* is an appropriate term for it. The time we spend scrolling and wandering down algorithmic rabbit trails is often time taken directly away from the nobler, if harder, tasks God has given us to do.

Order out of Chaos

What does seasonableness look like in action? Consider how it might shape one's habits in three different spheres: politics, relationships, and doctrine.

Political discourse is often held hostage by algorithms that colonize attention at the cost of thoughtfulness or even basic truthfulness. Suppose a particular political candidate articulates a position or belief in a public forum. Some people on social media quickly rally to extol that position or belief as obviously right, others quickly rally to decry it as obviously wrong, and each group uses calculated language to make clear to anyone on the fence that to be the "right kind of person," you must choose their conclusion. A seasonable response would be to log off social media, create distance between thinking and the algorithms, and study the candidate's beliefs carefully, comparing them to the truths of Scripture and established facts. This isn't a spineless call to equivocate; it's a call to pass judgment based on truth rather than artificial social pressure.

Suppose also a friend posts something that suggests he's struggling. These kinds of posts often generate a lot of feedback online, but much of this feedback begins and ends on the social media page. A seasonable response would be to reach out to this friend directly, preferably by phone or an in-person conversation, to care for him. The ease of simply replying to a post with a generic statement of sympathy belies the fact that such a place isn't where real care for one another happens. Pursuing embodied relationship places friendship where God designed it to be placed: in the real world.

Finally, suppose you are wrestling with a particularly difficult doctrine or trying to process a grievous instance of hurt at your church. Turning this struggle into public "content" on social media is tempting. You might sense that people would validate you via likes, comments, and shares. Yet social media, with its performative nature and superficial community, can't facilitate a clear head or clear heart. Seasonableness calls us to recognize how the allure of clout might muddy the vital process of wrestling with truth and processing our experiences. Dedicated time in Scripture, a deep dive into a book or catechism that has served the church for a long time, and a conversation with a trusted friend are seasonable responses that fit the significance of these situations.

The internet's dominance in our daily lives often leaves little space to think about it well. Using social media and smartphone technologies according to how their designers intend leads us into a numbing relationship with a never-ending stream of stuff. Our feeds alternate seamlessly between the useful, the tragic, the absurd, the outrageous, and the pointless. Constructing order out of this chaos requires the wisdom tradition of seasonableness, which both limits and orders our consumption. It's a way Christians can actively push back against the consuming, mind-numbing effects of "everything, all of the time."

There is no perfect, step-by-step plan for living wisely in the digital age. Biblical wisdom always aims to go beyond lists of rules and get inside: to make us certain kinds of people. Seasonableness is a small but real step toward becoming people who value and imitate the orderliness of our Creator, who caused light to shine out of darkness—both in the beginning and in our hearts (2 Cor. 4:6).

Discussion Questions

1. Throughout *Scrolling Ourselves to Death*, contributors (especially James) turn to wisdom and the Wisdom Literature of the Bible for guidance. Why is wisdom in particular so crucial when it comes to the technological revolution?

2. What is the critique summed up in the statement "Now . . . this"? Why was Postman so adamant about warning us of the consequences of a "Now . . . this" culture?

3. Consider and answer James's questions: "Do I know why I'm logging on right now? Is it to find something, share something, or accomplish a task? Or am I merely seeking distraction?" In your habits of using media, your phone, the web, or other apps, would you say most of your actions are passive or intentional?

4. Seasonableness, according to James, is a vital response to the problems we face in the digital world. When might it be a seasonable and reasonable time to use (a) your phone, (b) your television, and (c) your laptop or other screens. What might be unseasonable uses of these devices?

PART 2

PRACTICAL CHALLENGES
FACING CHRISTIAN
COMMUNICATORS

6

How the Medium Shapes
the Message for Preachers

Collin Hansen

TRAVEL BACK TO THE YEAR 1999. Less than half of American homes have an internet connection. Just about everyone is worried about the Y2K bug that could bring down the entire computerized infrastructure of banks, power plants, airlines, and more. Americans have been advised to prepare for the worst. Bathtubs fill with emergency water reserves.

In 1999, well-informed, concerned citizens get their news about Y2K or any other event from about twenty sources. They watch a national and local TV news program. Before dawn, they receive a newspaper delivery or two. They wait each week or month on a few magazine subscriptions. Their email inbox at work features some email forwards with bizarrely specific threats if you don't forward them yourself. (Nigerian princes appear to be prosperous yet need our help.) The average American tunes in during morning and evening commutes to a talk radio station or two. These sources—the daily paper, weekly copies of *Time* and *Newsweek*, the local TV mainstay, plus Tom Brokaw or Peter Jennings or Dan Rather —are widely shared and discussed by Americans from the neighborhood to the nation.

Jump ahead twenty years. The number of news sources available to Americans has expanded from twenty to two hundred . . . to two thousand . . . to two hundred thousand . . . to two million . . . to two hundred million . . . to two billion and beyond . . . to every person around the world who can open a Facebook profile, a couple burner X handles, an Instagram account for public and one to hide from the parents . . . and on and on.

It's an information revolution.

The codex was a revolution in the ancient world. The periodical was a revolution in industrial England and colonial America. The telegraph and radio shrank the world by accelerating the pace of news in the nineteenth and twentieth centuries. Television caused such a stir that Neil Postman wrote *Amusing Ourselves to Death*.

But nothing can compete with the information revolution unleashed by the internet toward the end of the twentieth century. A quarter of the way through the twenty-first century, we've only just begun to experience the effects. Especially since Apple's iPhone was introduced in 2007, the internet leaves almost nothing untouched. Walk into any modern classroom and you'll see the changes, with a shift away from up-front instruction toward hands-on collaboration.

Preaching, however, is a rare holdover from the world before 1999. The preacher still stands in front and commands our attention in a monologue that runs half an hour, give or take. Even though the practice may look similar, the experience of listening has dramatically changed along with just about everything else in the internet revolution. That change is the source of increasing concern and anxiety for preachers around the world. And it's a call to return to the foundations for why we preach in the first place.[1]

Primary Place of Formation

When preachers stepped into the pulpit twenty years ago, they held a knowledge advantage over most church members. Preachers knew more about the Bible, more about other Christians around the world, more about history and theology. That didn't mean the congregations would always agree with their pastors. They could read the Bible for themselves. They could purchase the history books from Borders or Amazon. They could subscribe to *Christianity Today*. They could watch Billy Graham on a TV special. This was the new evangelicalism of the twentieth century.

But this study required time, money, and effort for an ordinary Christian with commitments to work, family, and community. It was still a curated world, controlled by gatekeepers like *Christianity Today* editors Carl Henry and Kenneth Kantzer, publishers like Zondervan, or media-savvy producers like the Billy Graham Evangelistic Association. Like pastors, these gatekeepers benefited from broad agreement. TV shows and periodicals could sell more advertisements that way. Pastors could focus on study and shepherding while keeping one eye on the cable news and talk radio hosts most popular among their congregations.

The curated world has largely disappeared in the twenty-first century. The inconspicuous editor has been replaced by the opaque algorithm. And the algorithm knows more about us than any pastor

---

1 This chapter builds on Collin Hansen's "Discipled by Everyone and No One: Is the Internet Good for the Church?," Desiring God website, February 16, 2022, www.desiringgod.org.

or editor ever could. The algorithm gives us what we might not even admit we want—sometimes with access to our darkest secrets. By contrast, church leaders can give us only what they think we need. In a larger church, that sense might be generic. Even in a smaller church, the experienced preacher can make certain assumptions but largely knows about us what we're willing to volunteer. And that preacher might claim an hour at best per week for teaching. You-Tube, TikTok, and X occupy every waking moment we'll give them.

Compared to twenty years ago, the internet—not the local church—has become the primary place where Christians are formed today. Before their leaders ever speak, many church members already know what they believe. After all, they've been reading, listening, and watching their favorite teachers all week. And they expect their leaders to conform—or else. Preaching, then, is expected to confirm the convictions already developed through the internet.

No wonder so many church leaders feel like they've lost their footing this century. No wonder we've seen so much splintering around elections and news events. In the crowded internet, content creators seeking to grow their influence get attention by rushing to the extremes and making enemies. They rise by stepping on other Christians. Preachers can't—or at least shouldn't—do likewise. But then they suffer by comparison to the "courageous" voices seen on YouTube and heard on podcasts. In contrast to online influencers who show they "know what time it is" by their aggressive denunciations of other Christians, pastors in pulpits can seem like wimps. Unlike YouTube hosts, many pastors must answer to other church leaders or even the congregation. The bully pulpit of yesteryear is no match for unaffiliated, unaccountable voices whose bank accounts bulge with every new controversy that grows their subscriber numbers.

Does preaching even have a role, then, in the information age?

Unknown Outcome

Every preacher who confides in me thinks his situation is unique. Elders resign with accusations of theological drift. Younger members leave in frustration because pastors didn't change their sermon to speak about the latest viral video. Deacons break decades-long friendships after they discover a new favorite YouTube channel.

In the aftermath, preachers reflect on what they did wrong. Did they unintentionally offend someone? Should they develop a new policy for when to revise the pastoral prayer? Did their favorite person to quote do all the terrible things that podcast suggested? Why bother to prepare an original sermon each week when countless sermons by the world's best preachers are freely available online?

When one preacher feels this way, it's good to look in the mirror. When it's happening to one denomination, it's good to look at the culture of training leaders. When it's happening in every single church, it's a revolution.

Today, we're living through the early days of a revolution at least as dramatic as the Protestant Reformation and printing press. And we don't know the outcome.

Neil Postman can help us navigate this revolution. But first we need to understand how we got here—how the medium is the message, in the memorable phrasing of Marshall McLuhan, whose work influenced Postman.

In human terms, the Reformation wouldn't have been possible without the printing press. Martin Luther's Wittenberg became a publishing powerhouse to meet the unprecedented demand for his liberating message and supply of his ceaseless writings. Like X or YouTube in our day, the printing press made it possible for Luther to speak directly to the people even without the support of ecclesiastical authorities. This popular support buoyed his political protection. As a preacher, Luther depended on the support of German princes who guarded him from the emperor's armies, acting under orders of the pope who had condemned him as a heretic.

Even as a popular medium, accessible to the increasingly literate masses, the printing press still carried complex, dense theological arguments. In Postman's account, Protestant pastors such as Jonathan Edwards followed in this wake by carrying "tightly knit and closely reasoned expositions of theological doctrine" to popular audiences.[2] In the colonial America of Harvard, Yale, and Princeton, the ideal pastor was still a man of learning, of reason, of the Scriptures.

With the advent of television in the twentieth century, however, the ideal pastor was no longer expected to deliver difficult sermons. Postman identified televangelists such as Oral Roberts and Jimmy Swaggart, who made their riches through emotive appeals into a television camera. The medium wasn't designed or capable of transmitting the kinds of arguments made by Luther and Edwards. Television was designed to keep viewers engaged long enough to sell them soap via commercials. Postman wrote, "Christianity is a demanding and serious religion. When it is delivered as easy and amusing, it is another kind of religion altogether."[3]

In other words, television isn't a neutral medium for conveying the gospel message. In Postman's account, television necessarily truncates and trivializes the Christianity of Edwards and Luther and countless other teachers, going back to the apostles and Jesus himself. The television screen, Postman argued, "has a strong bias toward

2 Neil Postman, *Amusing Ourselves to Death*, 20th anniversary ed. (1985; repr., New York: Penguin Books, 2005), 54.

3 Postman, *Amusing Ourselves*, 121.

a psychology of secularism" and "wants you to remember that its imagery is always available for your amusement and pleasure."[4]

But if Postman found television too easy to turn off, too easy to give us what we want instead of what we need, what would he have concluded about the internet? Does the internet deliver popular access with complex arguments? Or does this medium—think of social media in particular—privilege certain messages that divide us?

How evangelicals answer these questions may determine whether their movement in the twenty-first century goes the way of Luther and Edwards or Roberts and Swaggart. Evangelicals have tended toward early adoption of technology—but, as Postman warned, not always to their benefit.

Early Adopters

American evangelicals in particular see technology as an indispensable aid in widespread, rapid communication of the gospel. They were quick to publish religious periodicals. They were quick to broadcast sermons on radio. They were quick to produce feature-length Bible films. They were quick to broadcast evangelistic meetings on television. And they were quick to recognize the fundraising power of these new media.

The same is true of the internet era. Evangelicals have been quick to post sermon audio and video to church websites. They have been quick to develop Bible apps that assist churches by featuring sermon texts, outlines, and illustrations. They have been quick to add campuses with recorded sermons instead of live, in-person preaching. They have been quick to develop multiverse churches for digital avatars. They have been quick to ask ChatGPT for help developing

---

4   Postman, *Amusing Ourselves*, 119, 120.

study questions for small groups. If you find a technologically savvy church, you almost certainly know it's evangelical and not Catholic, Orthodox, or mainline Protestant. When I began building the Gospel Coalition's internet publishing in 2010, I followed this example.

Sometimes early adoption by evangelicals lacks theological reflection and spiritual discernment. Each new medium opens new possibilities—but also takes them away. Postman wrote, "Each medium, like language itself, makes possible a unique mode of discourse by providing a new orientation for thought, for expression, for sensibility."[5]

Evangelicals tend to emphasize what the new technology adds but not always recognize what it takes away. More than a decade ago, I published Matthew Barrett's gently critical essay that argued why preachers should still use physical Bibles instead of iPads in the pulpit.[6] Barrett observed that a worn Bible communicates care, longevity, and attention. It won't be confused for anything else. But an iPad could suggest just about anything and distract the congregation. He wrote,

> The sight of an iPad screams instant access to *Sesame Street* on Netflix. For the adult, the tablet is an immediate window into his or her social life. As advertised, the iPad is *ESPN Magazine*, a Visa card statement, decorating ideas on Pinterest, hotel reservations in Hawaii, the latest college football scores, Adele on iTunes, directions to the nearest Starbucks, instant tracking of the stock market, and, oh yes, the Bible, alongside thousands of your favorite e-books.[7]

---

5 Postman, *Amusing Ourselves*, 10.
6 Matthew Barrett, "Dear Pastor, Bring Your Bible to Church," The Gospel Coalition, August 18, 2013, https://www.thegospelcoalition.org/.
7 Barrett, "Dear Pastor."

In short, the device distracts us from focusing on God's word. When we published the article in 2013, smartphones hadn't yet achieved near-universal adoption, especially among youth. But the arguments would apply even more to a congregation where each person can toggle between the Bible and their text messages, social media notifications, and news apps. Subsequent studies have shown a precipitous decline in reading proficiency among fifteen-year-olds—across the developed world—starting in 2012.[8] The most likely culprit? The introduction of smartphones in educational spaces. Likewise, the COVID-shutdown educational declines can be attributed in large part to internet distractions in the home learning environment. While Barrett's 2013 article speculated about potential problems, studies since have confirmed their truth.

Back in 2013, however, his article received overwhelming backlash from online commenters. I had never seen such an intense wave of online criticism. Why? Because American evangelicals largely equate technology with progress. Arguing against technology is rejecting the Spirit's movement in our day. If someone finds Jesus in the multiverse, who are we to judge? If video-venue campuses are growing, they must be doing something right. And if livestreamed church is more convenient for a busy family, why would we make them feel guilty for not attending in person? If an iPad makes preaching easier, then iPads must be superior to flipping around a print Bible to find some obscure book like Micah.

As a publisher who works primarily to produce digital resources for the church, I'm not easily categorized as a Luddite. If I thought digital technology was bad for Christians, I wouldn't invest most of my life in recording videos and podcasts and writing newsletters

---

8 Derek Thompson, "It Sure Looks Like Phones Are Making Students Dumber," *Atlantic*, December 19, 2023, https://www.theatlantic.com.

for digital distribution. However, I strongly believe Christians must heed the warnings of Postman and others by recognizing that technology gives *and* takes away. We must discern the good as well as the bad. And we shouldn't assume the positive will always outweigh the negative. Luther and Edwards could combine complicated arguments with broad appeal. Where do you see patient, generous consideration of theological reasoning on the internet? Where you find it, praise God. And where you don't, consider how the medium tempts us to truncate a gospel message that runs as deep as it flows wide.

Not every church should attempt to produce the same kinds of resources we publish at the Gospel Coalition. In fact, most churches will excel in focusing on the ministry we can never duplicate as a digital parachurch. We can produce resources that bless the masses. These will often be used by the Holy Spirit to encourage and challenge individuals. But the world can do without what we do. The world can't do without the local church. And the world can't do without preachers who exposit God's word for people they know and love by name.

Think Small

I urge preachers to think small. My recommendation goes against our instincts and sometimes our temptations. Before radio, television, and the internet, even the best preachers rarely became known beyond the reach of their voice. Exceptions such as George Whitefield and Charles Spurgeon prove the rule. Most preachers never reached anyone outside their church walls.

With the authority crisis facilitated by the digital revolution, preachers may be tempted to fight fire with fire. Start the podcast. Share the sermon clips. Expand the campuses. Launch the newsletter.

I believe some have been called for exactly that purpose, and I'm glad to partner with many of them. They are the exception, however. Creating digital content sometimes feels like hand-pumping water from a river into the ocean. You're adding to the volume, but it's hard to notice any difference. Every moment online is a flood of voices vying for attention—voices aimed at recipients who are never more than an unseen audience, faceless "fans," and accumulated metrics (downloads, likes, views).

By contrast, what about a voice that knows your name, from a preacher who can name your pain? That's like finding an oasis in the desert. In the information age, most preachers should think small. I know some who won't even record and post their sermons. These messages belong to a particular time and place and people. They belong to the depths of a specific context. They don't bob along the surface of the vast content ocean.

When we preach to a particular people, place, and time, we may find fewer problems with declining attention spans and increasing distractions. What if Christians knew this message was for them—right here, right now? What if they knew they couldn't get that message unless they showed up in person? What if they knew it wasn't interchangeable with a digital devotional or sermon podcast or YouTube clip?

Even campuses with recorded sermons don't prepackage the music. They recognize the congregational experience can't be replaced. You can't perform a digital baptism. You need actual water. You can't partake of the Lord's Supper in the multiverse. You must taste the bread and wine for yourself.

Ask anyone who preached during COVID-19 lockdowns, and they will tell you: it's not the same when you preach into a camera set up in an empty room as when you talk to your people sitting shoulder to shoulder. Some traditions demand verbal feedback as pastors preach. Many preachers feel the Spirit's nudge as they pick up nonverbal feedback in the congregation. They might focus or shift their address when they see familiar faces and know the story behind their joy or grief. Sometimes a sermon that resonates in one venue flops in another. Preaching doesn't feel like a studio sitcom recorded with a laugh track. It works best with a live studio audience.

Itching Ears

In the information age, authority has been vested in users holding the smartphone. And the algorithm serves us what we crave, not necessarily what we need for spiritual nourishment. We read from Paul in 2 Timothy 4:3–4: "For the time is coming when people will not endure sound teaching, but having itching ears they will accumulate for themselves teachers to suit their own passions, and will turn away from listening to the truth and wander off into myths." That time has come with the digital revolution. Everyone can customize a lineup of teachers who suits his or her passions, regardless of the truth. And algorithms beckon itching ears (and scrolling eyes) to wander off into myths.

It might seem like the wrong season, then, to produce bespoke sermons for a specific people, time, and place. But the command from Paul to Timothy, and then to preachers in every subsequent age, remains the same: "Preach the word; be ready in season and out of season; reprove, rebuke, and exhort, with complete patience and teaching" (2 Tim. 4:2).

Preaching to a camera might be more efficient for reaching a mass audience. And yet Paul never set target reach or engagement metrics for his protégé Timothy. The inefficiency of real people in an actual place at a specific time seems to help establish and confirm the legitimacy of our message. Paul tells Timothy, "As for you, always be sober-minded, endure suffering, do the work of an evangelist, fulfill your ministry" (2 Tim. 4:5).

Suffering doesn't translate well across the algorithm for digital content creators. We don't see them in three dimensions. They exist to scratch our itches, to fulfill our wishes. They don't possess authority that we don't grant them.

But what if that preacher knew the worst about us? What if we knew how that preacher suffered for the gospel's sake? What if we'd seen the sobriety of his behavior? Would we, then, be more willing to accept his rebuke? If we knew of his faithful endurance, would we heed his reproof?

Digital media reshapes the sermon by putting listeners in the position of authority and serving up content that confirms what they want to hear. And while some of that content may align with Scripture, we're tempted not to seek out and heed the full counsel of God. Whatever doesn't interest us, or perhaps even offends us, can be skipped with the tap or swipe of the finger. Even the most edifying online content is curated according to our perceived interests and needs. But a sermon set by continuous exposition through the Bible or a liturgical calendar reminds us not to center our felt needs as if we're searching on Google. It's striking how often a sermon we didn't plan from a passage we didn't choose becomes a "word in season" (Prov. 15:23).

We can't trust our hearts to determine what we want or need. But we can trust the one-way voice of God found in the Scriptures, all the way from Genesis to Revelation. And we need preachers who will proclaim that message, in season and out. The best preaching tells us what we don't always want to hear. The best preachers, committed to teaching us to obey everything Jesus commanded (Matt. 28:20), don't abandon us to the algorithm. Let's encourage them by likewise giving ourselves, in private devotion and public proclamation, to the ministry of the word.

Discussion Questions

1. Why is preaching a communication form that has remained (relatively) untouched in the last decade when compared to other forms of communication? Envision a future fifty years from now in the United States. Will preaching still be around? *Should* preaching still be around?

2. Reflect on Hansen's comments about the local church. Identify some activities of the local church that cannot be replicated online. How might this shift our perspective on how the local church should expend most of its energy?

3. More and more of our friends are opting for podcasts or videos in place of in-person sermons. Reflecting on the points made in this chapter, how would you explain to a friend why it is vital for him or her to attend to the word of God in person?

4. Many pastors feel the weight of being compared to some of the most talented orators in the world. Carve out a few minutes to write a note or send a text telling your pastor about something the Holy Spirit did in your heart through his preaching. Think of other ways you might encourage your pastor.

7

Apologetics in a Post-Logic World

Keith Plummer

"WHAT IS TRUTH?"

Multitudes of twenty-first century people daily echo the infamous query of a first-century Roman governor (John 18:38). They are wanderers in our post-truth world. For various reasons—from the assumption that truth claims are really veiled attempts to exploit or oppress to the myriad of competing, often contradictory, claims vying for our severely limited and increasingly diminished attention—many of those we seek to evangelize are wary and weary of yet another argument for *the* truth.

Some have called our current moment an "epistemological crisis."[1] And for Christians, any epistemological crisis is also a crisis of mission. How do we win people over to the gospel's truth if the very concept of solid truth is jeopardized?

---

1 See, for example, Brett McCracken, "How to Weather the Worsening Trust Crisis," The Gospel Coalition, February 2, 2021, https://www.thegospelcoalition.org.

Here's how Bonnie Kristian describes the situation: "Our information environment is chaotic and overwhelming, rife with conspiracy theories, 'fake news,' and habit-forming digital manipulation. It is breaking our brains, polluting our politics, and corrupting Christian community. It may be the most pressing and unprecedented challenge of discipleship in the American church."[2]

A discipleship challenge, absolutely. But this environment presents a formidable challenge to the tasks of evangelism and apologetics as well.

To a large extent, though not exclusively, technological change has contributed to the current crisis. This is something Postman foresaw. In 1985, he called television our culture's principal mode of knowing about itself and the model for how the world was to be staged.[3] Today, internet-connected smartphones, laptops, and tablets hold those designations.

Postman's insights from forty years ago are helpful for understanding how today's information and communication technologies have contributed to epistemological breakdown. It's an epistemological crisis marked by incoherence, inattentiveness, and meaninglessness—all standing as obstacles to communicating the gospel. Though challenging, they're not insurmountable. In the biblical story we have the resources for resisting and engaging a culture in epistemological free fall.

Technology Changes How We Think

The Christian faith makes objective truth claims about the triune God's identity and activity, as well as about the world he created

---

2 Bonnie Kristian, *Untrustworthy: The Knowledge Crisis Breaking Our Brains, Polluting Our Politics, and Corrupting Christian Community* (Grand Rapids, MI: Brazos), 1.

3 Neil Postman, *Amusing Ourselves to Death*, 20th anniversary ed. (1985; repr., New York: Penguin Books, 2005), 92.

and sustains. Apologetics involves offering reasons for accepting these claims as worthy of consideration and trust. If, as Postman contended, dominant communication technologies form the intellectual faculties of a society's members and, to a significant extent, influence how people think and converse, then an important part of Christian apologetics is understanding those thought patterns and the obstacles they present.

Communication technologies and media play a crucial role in influencing a culture's plausibility structures. Felicia Wu Song, a cultural sociologist of media and digital technologies, notes, "Though technological affordances certainly do not *determine* behavior, it is important to recognize how they create environments that make some worlds and behaviors more imaginable and achievable than others."[4]

Postman's foremost concern in *Amusing Ourselves to Death* is revealed in the book's subtitle: *Public Discourse in the Age of Show Business*. Postman argued that the age of the image or the "Age of Show Business" represented a revolutionary contrast to the typographic age or the "Age of Exposition," in which the written word was central to societal conversation.[5] Postman was convinced "that the media of communication available to a culture are a dominant influence on the formation of the culture's intellectual and social preoccupations."[6]

This in turn, affects the nature of public discourse. The introduction of a new communication medium does this "by encouraging certain uses of the intellect, by favoring certain definitions

---

4   Felicia Wu Song, *Restless Devices: Recovering Personhood, Presence, and Place in the Digital Age* (Downers Grove, IL: IVP Academic), 26.

5   Postman explains that he uses conversation "metaphorically to refer not only to speech but to all techniques and technologies that permit people of a particular culture to exchange messages." See Postman, *Amusing Ourselves*, 6.

6   Postman, *Amusing Ourselves*, 9.

of intelligence and wisdom, and by demanding a certain kind of content—in a phrase, by creating new forms of truth-telling."[7] Postman regarded the epistemology created by television as inferior to print-based epistemology, calling the former "dangerous and absurdist."[8] In print, discourse was characterized by coherence, seriousness, and rationality.

To illustrate the features of the print-oriented "typographic mind," Postman pointed to the popular nineteenth-century debates between Abraham Lincoln and Stephen Douglas (when neither man was a presidential candidate), one of which lasted seven hours. Postman noted this was not exceptional for the period. People were generally able to follow and appreciate complex arguments and rhetorical sophistication because "both the speakers and their audience were habituated to a kind of oratory that may be described as literary."[9] Another characteristic of the typographic mind was the ability to attend to discourse for an extended period. Postman asked the reader, "Is there any audience of Americans today who could endure seven hours of talk? Or five? Or three? Especially without pictures of any kind?"[10] He was asking this in 1985. Forty years later, we must now ask about attention in terms of not hours but minutes.

Our Scrolling-Shaped Minds

Postman regarded the television age as synonymous with the age of entertainment, because entertainment is what television does best. Anything televised must accommodate television's ideology

---

7   Postman, *Amusing Ourselves*, 27.

8   Postman, *Amusing Ourselves*, 27.

9   Postman, *Amusing Ourselves*, 48.

10 Postman, *Amusing Ourselves*, 45.

or bias, which Postman argued ran contrary to what characterized typographic discourse: the "sophisticated ability to think conceptually, deductively and sequentially; a high valuation of reason and order; an abhorrence of contradiction; a large capacity for detachment and objectivity; and a tolerance for delayed response."[11] Reading requires thoughtfulness, attentiveness, reflection, and the ability to follow a train of thought or line of reasoning carefully. Comprehension presupposes coherence of the text to which one is attending.

Television, Postman argued, fosters incoherence. By receiving an array of images, programs, commercials, and news items detached from larger contexts, explanations, or analysis, the sort of mind cultivated by TV was less attentive, less reflective, less able to connect the dots. Postman also argued that "the telegraphic person values speed, not introspection."[12]

As a technology, the internet has exacerbated all these tendencies. Postman was already observing shorter attention spans and eroded capacities for logic in the TV-shaped world of the 1980s. These troubling trends have grown considerably worse in the world shaped by scrolling feeds and social media.

Postman's focus on incoherence is of particular interest to our exploration of the current epistemological crisis. I frequently meet suspicion, if not skepticism, when I talk with Christians about the value and necessity of reasoning with people about the claims of the faith. It's not uncommon for people to object: "People don't care about logic. As a matter of fact, many seem quite comfortable with contradiction."

11 Postman, *Amusing Ourselves*, 63.
12 Neil Postman, "Five Things We Need to Know" (talk, Denver, CO, March 28, 1998), https://web.cs.ucdavis.edu/~rogaway/classes/188/materials/postman.pdf.

Part of the reason contemporary people have become comfortable with contradiction is that they've been shaped by a scrolling world in which discontinuity reigns. "And in a world of discontinuities, contradiction is useless as a test of truth or merit, because contradiction does not exist."[13] Postman told a story that illustrated this in his own day. He tried to show his students that they had asserted two opposite arguments in the same paper. Which of the two did they mean?

> They are polite, and wish to please, but they are as baffled by the question as I am by the response. "I know," they will say, "but that is *there* and this is *here*." The difference between us is that I assume "there" and "here," "now" and "then," one paragraph and the next to be connected, to be continuous, to be part of the same coherent world of thought. That is the way of typographic discourse, and typography is the universe I'm "coming from," as they say. But they are coming from a different universe of discourse altogether: the "Now . . . this" world of television.[14]

Postman's vignette brings a smile of familiarity to my face. I frequently read papers in which students cite wildly contradictory passages from different sources, failing to realize they stand in sharp disagreement. This is in part due to such factors as their being hurried to complete the assignment and their desire to meet the required page limit and minimum number of reference works. In most cases, however, it's also because they were preoccupied with a host of distractions—texts, notifications, and social media—while

13 Postman, *Amusing Ourselves*, 110.
14 Postman, *Amusing Ourselves*, 110.

writing. Distraction·has become a norm for them, and it manifests in the disjointed reasoning that often defines their writing.

But it is not just young students. It is the majority of us, regardless of our age. We are members of what Joel Nigg, an expert on children's attention problems, calls "an attentional pathogenic culture."[15] The velocity of the "Now . . . this" world has increased exponentially, with the result being diminished attention and less ability to think in wholes.

Consider how frequently and rapidly in the course of any twenty-minute period we shift our attention from one thing to another. We scroll through various apps and task-switch constantly from one browser window to another. We read and try to reply to emails having nothing to do with each other. We usually have multiple screens on, with countless apps open, and our attention constantly toggles between it all. It's no wonder our capacity to think coherently is severely atrophied.

Postman's observation in 1985 is startlingly applicable to life in the scrolling age: "My point is that we are by now so thoroughly adjusted to the 'Now . . . this' world of news—a world of fragments, where events stand alone, stripped of any connection to the past, or to the future, or to other events—that all assumptions of coherence have vanished."[16]

Postman was not saying contradictions ceased to exist. We have just become less able to apprehend them. Writing about the internet's atomization of our thoughts, Nicholas Carr observes, "A search engine often draws our attention to a particular snippet of text, a few words or sentences that have strong relevance to whatever we're

---

15 Johann Hari, *Stolen Focus: Why You Can't Pay Attention—And How to Think Deeply Again* (New York: Crown, 2022), 11.

16 Postman, *Amusing Ourselves*, 110.

searching for at the moment, while providing little incentive for taking in the work as a whole. We don't see the forest when we search the Web. We don't even see the trees. We see twigs and leaves."[17]

While words proliferate on the internet, they are situated in a venue designed to grab attention rather than spark careful reflection—to provide quick, utilitarian answers rather than stimulate deep thinking in pursuit of wisdom.

Evangelism to the Scrolling-Shaped Mind

What can evangelists and apologists do to effectively communicate the gospel to internet-shaped people who are increasingly unreflective, uncritical thinkers with short attention spans?

To start, we should be careful that our methods don't simply exacerbate the problem. Gospel proclaimers need to go where the unreached are. But this doesn't mean we must use the communicative styles and practices of those we are trying to reach. "Meeting people where they are" does not necessarily mean "meeting people in the media formats they most prefer." What if some media formats actually work against the comprehension of the truths we're trying to communicate? Young people are on TikTok, but we must ask ourselves whether TikTok's form is well suited to theological reflection and gospel proclamation.

We should also prioritize embodied, relational contexts for evangelism and apologetics. Even though we are connected in a manner unlike any other in human history, loneliness is at an all-time high. This is an opportunity for Christians to show digitally distanced people what true connection is like. Caleb Wait argues that it might "be better for the church's imagination to be taken up less

17 Nicholas Carr, *The Shallows: What the Internet Is Doing to Our Brains* (New York: W. W. Norton & Co, 2010), 91.

with the efficiency and reach of digital media and communication for sharing our message, and taken up more with the heart and good cheer of in-person discussion and debate."[18] I think he's on to something. This isn't to say that there is no place for using technology for evangelistic and apologetic purposes. But we need to think about how we can use such media as invitations to face-to-face encounters where people can ask questions (and have them answered) and witness Christian community.

The hurried impatience of the scrolling world is a big part of why reflective, careful thinking is on the decline. Fast, ephemeral technology naturally conditions us to be fast, ephemeral thinkers. A thought comes into our mind briefly, but then something else grabs our attention and our thoughts go in a different direction, and so forth. Our thinking is as all over the place as our Facebook feeds; our thoughts come and go too fast to be truly considered on a substantive level. To be effective evangelists and apologists in a fast-thinking world like this, we need to model a slower, more reflective mode of thinking through important topics. Instead of making arguments about God or theology via back-and-forths on social media or in YouTube comments sections, we should nudge people toward the importance of slower, more deliberate, more focused thinking about topics of eternal significance.

Postman's concerns about the destructive effects of the age of the image on our ability to reason are certainly relevant to our contemporary apologetic task. However, we must resist treating people in a reductionist manner, as if they're only reasoning creatures. Our intellects suffer from our frenetic, media-saturated habits—but so do our imaginations. Kevin Vanhoozer contends we need to expand

---

18 Caleb Wait, "The Medium Is the Mania: Anxiety as a Feature, Not a Bug, of Digital Media," *Modern Reformation*, July 1, 2023, https://www.modernreformation.org/.

our conception of cognitive faculties to include the imagination, which he describes as "the power of synoptic vision: the ability to synthesize heterogeneous elements into a unified whole. *The imagination is that cognitive faculty that allows us to see as whole what those who lack imagination see only as unrelated parts.*"[19] He notes, "The point of narrative is not merely to assert 'this happened, and then this happened.' Narratives make another kind of claim altogether: 'look at the world like this.' "[20]

Narratives then, do more than inform by way of propositions and arguments. They call us to see and inhabit the world in particular ways. To grasp this, think of a popular storytelling form like movies. M. Night Shyamalan's *Signs* (one of my favorites), for example, doesn't just tell a story but aims for viewers to see life as providentially ordered by some kind of higher power and not as an array of random, purposeless events. Narratives, whether literary or cinematic, communicate worldviews.

The grand, glorious biblical story is what minds trained to think in fragments need in order to see how the pieces of human experience fit together in a larger whole. Evangelism and apologetics are about proclaiming and defending a sweeping, coherent, and imagination-gripping story about the nature of reality that explains our deepest desires as well as the reasons for our resistance to the truth.

The Christian faith has propositional content, and thus reasoning and argumentation are called for. But though our theology should be solidly grounded in the Bible, God didn't give his message via the medium of a systematic theology. We must keep this in mind

---

19 Kevin Vanhoozer, *The Drama of Doctrine: A Canonical Linguistic Approach to Christian Theology* (Louisville, KY: Westminster John Knox Press, 2005), 281.

20 Vanhoozer, *Drama of Doctrine*, 282.

as we strive to introduce people to the Savior. Scott Christensen puts it well: "The drama of story (and poetry) adds sound, color, flavor, texture, and aroma to revelational truth that simply cannot be conveyed by theological abstraction. *Drama* and *doctrine* are equally necessary and complementary components to our understanding of the Christian faith."[21]

We must not give up on rationality. Though contemporary people might seem fine with a "post-logic world," we know that, as creatures made in the image of God, rationality is an unavoidable part of who we are. Despite the suppression of the truth or attempts to fully relativize truth ("You do you"), people can't completely flee from rationality, arguments, and logical defenses. They can't help but think in certain ways, and they'll betray this whenever they offer logical arguments and "evidence" for, let's say, their disbelief in the authority of Scripture (like pointing to the Bible's alleged contradictions).

To give up on reasoning because people are unreasonable is equivalent to ceasing to be moral because people are immoral or don't acknowledge objective moral norms. We must keep trying to engage people in deeper thinking about topics that matter—God, the gospel, and how to live in alignment with God's word.

The meager reasoning skills, fleeting attention, and continual distraction we encounter might be frustrating at times. But we shouldn't despair. The gospel's truth remains simple, beautiful, and powerfully transformative. Let's keep communicating the truth as clearly and beautifully as we can, trusting that God will soften scrolling hearts and open distracted minds so the message is clearly received and, by his grace, trusted.

---

21 Scott Christensen, *What About Evil? A Defense of God's Sovereign Glory* (Phillipsburg, NJ: P&R, 2020), 252.

Discussion Questions

1. If "drama and doctrine are equally necessary" to understanding Christianity, then a lack of either will leave us ill-informed. How does modern technology undermine both doctrine and drama? Are there uses of modern technologies that could increase our grasp of Christian doctrine and drama?

2. In your experiences with sharing the gospel or defending aspects of the Christian faith, what are the most difficult parts of "arguing persuasively" (Acts 19:8 NIV) in a post-logic world? What have you found to be effective in evangelism or apologetics in a world like this?

3. Even if you haven't read *Amusing Ourselves to Death*, it's probably clear by now that Postman believed "image-based" communication is inferior to "print-based" communication and tends to erode our capacities to think critically. Do you agree with this assessment? What are the implications of an image-based society for a word-based faith like Christianity?

4. Quoting Postman, Plummer argues that technology changes our minds by "favoring certain definitions of intelligence." What types of intelligence are favored by dopamine technology? How does the elevation of these kinds of intelligence change our thinking as a society and create challenges for sharing the gospel?

8

Telling the Truth about Jesus in an Age of Incoherence

Thaddeus Williams

THE AGE OF FAITH, the age of reason, and the age of science. Some ages in human history are named for their dominant epistemologies, their understanding of how we best come to know what is real.[1] Do we *tradition* our way to truth, *exegete* our way to truth, *autonomously reason* our way to truth, *science* our way to truth, *feel* our way to truth, *entertain* our way to truth, or *scroll* our way to truth? Other ages take their names from their breakthrough technologies: the Stone, Bronze, and Iron Ages with the advance

---

1   Every major cultural shift in the history of Western culture was an epistemological shift before (and while) it was a cultural shift. The sixteenth-century Protestant Reformation, was, among other things, an epistemological shift away from papal declaration and tradition back to the fount or source (*ad fontes*) of all truth—the Bible alone. The eighteenth-century age of reason, Enlightenment, and French Revolution (which quickly devolved into the Reign of Terror) were also epistemological shifts toward the unquestionable Cartesian self (René Descartes) as the starting point of knowledge.

of new, increasingly durable tools and weaponry; the industrial age with power-driven machines displacing hand tools; the atomic age with the unleashing of nuclear weapons; the space age with manned projectiles fired to the moon and back, and so on.

We have ages named for their epistemologies and ages named for their technologies. Which did Neil Postman have in mind when he coined the "Age of Show Business" in the subtitle of *Amusing Ourselves to Death?* Is it an epistemological or a technological moniker?

The answer is yes. And this is one of Postman's most profound points. On one level, "show business" can be thought of as a technological breakthrough, harnessing electromagnetic waves, cathode ray tubes, and other sophisticated components to bring recorded images and sounds to life before our eyes. But show business, for Postman, simultaneously represented an epistemology. No technology is epistemologically neutral. Each will foster certain habits of the mind, for better or for worse, while rendering other ways of knowing irrelevant and favoring certain views of reality over others.[2]

According to Postman, the printing press's speedy proliferation of bound books (eight million in its first fifty years) helped Westerners form typographic culture, with its own set of epistemic habits. Reading "encourages rationality."[3] It invites us to "follow a

---

2   As a contemporary case in point of Postman's observation, Bo Burnham in his dark Netflix comedy special *Inside* notes the real world used to be the *real* world: the world outside filled with beaches, mountains, sunsets, ladybugs, architecture, hiking trails, and the like. The virtual world of social media has since become the so-called real world, as the flowers become something to pluck and put over one's eyelids to get that great likeworthy Instagram shot, the beautiful sunset skyline becomes a mere backdrop before which to pose with a duck face. The actual world becomes a mere prop for what becomes our functional real world: the world of social media.

3   Neil Postman, *Amusing Ourselves to Death*, 20th anniversary ed. (1985; repr., New York: Penguin Books, 2005), 26, 51.

line of thought, which requires considerable powers in classifying, inference-making and reasoning."[4]

The printing press and its corresponding epistemological virtues have been largely swept aside in the age of show business with the "Now . . . this" phenomenon of television. Postman clarified, "the fundamental assumption . . . is not coherence but discontinuity. And in a world of discontinuities, contradiction is useless as a test of truth or merit, because contradiction does not exist."[5] Under the spell of show business, as both a technology and an epistemology, we have become a public "adjusted to incoherence" and "amused into indifference."[6]

Such passages from Postman transport me to a plane somewhere over the Pacific, destined for Kathmandu, Nepal, on a mission trip some twenty years ago. It was the first time I read Francis Schaeffer—in some ways the Christian version of Neil Postman. At the opening of *The God Who Is There*, Schaeffer tries to explain the generation gap between the 1960s counterculture and their parents. It came down to what Schaeffer dubs "antithesis." He writes,

> [The last generation] took it for granted that if anything was true, the opposite was false. In morality, if one thing was right, its opposite was wrong. This little formula, "A is A" and "if you

---

4 Postman, *Amusing Ourselves*, 51.

5 Postman, *Amusing Ourselves*, 110.

6 Postman, *Amusing Ourselves*, 110–11. Recent research into the online habits of Gen Z shows how much digital technology has shaped epistemology toward "incoherence" and "indifference." Summarizing the research, one writer said, "Gen Zers know the difference between rock-solid news and AI-generated memes. They just don't care. . . . Where older generations are out there struggling to fact-check information and cite sources, Gen Zers don't even bother. They just read the headlines and then speed-scroll to the comments, to see what everyone else says." Adam Rogers, "The Secret Digital Behaviors of Gen Z," *Business Insider*, June 25, 2024, https://www.businessinsider.com.

have A, it is not non-A," is the first move in classical logic. If you understand the extent to which this no longer holds sway, you will understand our present situation. Absolutes imply antithesis.[7]

Seventeen years later, Postman echoed him: "Mutually exclusive assertions cannot possibly both, in the same context, be true. . . . My point is that we are by now so thoroughly adjusted to the 'Now . . . this' world of news . . . that all assumptions of coherence have vanished."[8]

How then, do we tell Christian truth to people in the age of show business—and now the age of scrolling—who deny antithesis, don't seem to mind (or notice) incoherence, and believe contradiction doesn't exist? Here are four potential strategies for Christian truth-telling in a post-logic world.

Beyond Relativization

To minds conditioned by relativization—the loss of logical consistency—we declare that Jesus is *the* way, not one of many equal ways.

On that mission trip to Nepal, my fellow missionaries and I were baffled when we found immediate but superficial success. We would share Jesus with Hindus in the streets of Kathmandu and the typical response was "Great, we accept." It took us too long to realize the epistemological disconnect. We did not realize until after more intentional listening that many of our new Hindu friends believed in several million deities, so adding one more to the pantheon was hardly a stretch. Plus, much like the rejection of antithesis and vanishing of coherence that Schaeffer and Postman

---

7 Francis Schaeffer, *The God Who Is There* (Downers Grove, IL: InterVarsity, 2020), 2–3.
8 Postman, *Amusing Ourselves*, 109, 110.

lamented in the West, many Eastern traditions are unconcerned with (and in some cases even celebrate) contradiction.

East or West, we must make it clear when we present the good news of Jesus that we're not offering Christ as *a* way, equal among all others, but as *the* way, unparalleled in his goodness, trustworthiness, grace, historical reality, bodily resurrection, and saving power.

It's at this point of exclusivity where too many Christians turn mealymouthed and shifty. With perhaps noble intentions to reach those "adjusted to incoherence," we reduce the gospel itself to incoherence. Nearly half of Christian millennials believe "it is wrong to share one's personal beliefs with someone of a different faith in hopes that they will one day share the same faith."[9] This turns the Great Commission into the Not-So-Great Suggestion. Forty percent affirm that "if someone disagrees with you, it means they're judging you,"[10] a notion tied to today's dogma that love requires us to accept and even celebrate others' beliefs and behaviors.[11]

For fear of becoming social pariahs, marked with scarlet *B*s for bigotry, many make a fatal missiological move. They relativize the gospel, presenting it (if at all) as *a* truth, or *my* truth, but never what it is—*the* truth. Believing in Jesus as opposed to, say, Buddha, Joseph Smith, or even Joseph Stalin is put on the same level as preferring mint chocolate chip over cookie dough or rocky road.[12]

---

9 "Almost Half of Practicing Christian Millennials Say Evangelism Is Wrong," Barna, February 5, 2019, https://www.barna.com/.

10 "Almost Half of Practicing Christian Millennials."

11 See my article "Disagreement Equals Hate? The False Assumption Sweeping the Nation," *World*, February 8, 2022, https://wng.org/.

12 Sociologist Thomas Luckmann noticed this rising trend back in the 1960s. "The individual," Luckmann explains, "is left to his own devices in choosing goods and services, friends,

The missionary (if he can be called that) in this exchange can expect the familiar retort, "That's true for you, but not for me, and I'm glad you found something that works for you." This is precisely the weak answer such a weak gospel elicits.

A church in my neighborhood makes a point every Sunday to declare from the stage, "We do not have all the answers." A leader at this church explained to me that they don't want to come off as know-it-alls or turn away folks who don't like religious dogmatism. But at a certain point, we must join the courageous Christians of history to declare without apology or qualification, "Jesus Christ is *the* answer! You can be saved from your sin by God's grace alone through faith alone in Christ alone."

Such bold, Christ-centered, and exclusive dogmatism does carry social consequences. But if our spiritual forebears could endure the hungry beasts of the Roman Colosseum, beheading, imprisonment, impalement, the gulags, and other agonies for the sake of telling the truth about Jesus, we can certainly cope with being called narrow-minded or intolerant.

When the gospel went from Jerusalem to Judea and Samaria all the way to Rome in the first century, it was not because Christians believed "it is wrong to share one's personal beliefs with someone of a different faith in hopes that they will one day share the same faith." The same goes for the costly missional successes in hostile places like Iran and China today. The good news doesn't advance through those who relativize their faith to fit the surrounding culture's incoherence. The gospel advances through Christians

---

marriage partners, neighbors, hobbies and . . . even 'ultimate' meanings in a relatively autonomous fashion. The consumer orientation, in short, is not limited to economic products but characterizes the relation of the individual to the entire culture." Luckman, *The Invisible Religion: The Problem of Religion in Modern Society* (New York: Macmillan, 1967), 98.

with the boldness to assert the *antithesis* (Schaeffer's term) and the *contradiction* (Postman's) between Christ *who can save* and Mithra, the Caesar, Allah, and communist governments *who cannot save*.

In an age of scrolling, an age of relativizing truth, we must be clear that Jesus is not just another lifestyle choice or guru to emulate; he is "the same yesterday and today and forever" (Heb. 13:8) and "*the* way, *the* truth, and *the* life" (John 14:6).

Beyond Trivialization

Postman lamented how the age of show business tended to strip the weightiness out of weighty things—reducing everything to innocuous, lightweight entertainment. To minds conditioned by trivialization—the loss of sacredness—we present Jesus as glorious Lord, not as another profane or glitzy product.

This has been amplified a millionfold in the age of scrolling, with the shift from television to smartphone technology. Within a sixty-second scrolling session, the social media "user"[13] may witness chilling scenes of human carnage in Ukraine, corpses in Israel, a groomsman losing his lunch at the altar, and an orange kitten that looks like Donald Trump.

In the 1990s, Postman marveled at the fact that 260,000 billboards lined America's highways. Americans took an estimated 41 million photographs per day while American homes contained 460 million televisions and 60 billion pieces of advertising junk mail littered their mailboxes each year.[14] Research from the latter half of the 1990s found the typical American was exposed to sixteen

13 In the Netflix documentary *The Social Dilemma* (2020), Edward Tufte notes, "There are only two industries that call their customers 'users': illegal drugs and social media."

14 From a talk given by Neil Postman on July 28, 1993, posted online as "Neil Postman Talk in LA 1993/7/28 (VPRI-0131)," Yoshiki Oshima, April 20, 2016, YouTube video, https://www.youtube.com/.

thousand advertisements over the course of a twenty-four-hour period, factoring to more than six million advertisements a year.[15] All that outrageous information overload was before Steve Jobs unveiled the iPhone in 2007, before the smartphone revolution and the explosion of social media. The age of scrolling, far more than the age of show business, bombards our consciousness with an overabundance of content. When confronting the tech geniuses at Apple in the 1990s, Postman observed, "We are glutted with information. We are drowning in information, have no control over it and don't know what to do with it. We no longer have a coherent conception of ourselves and our universe and our relation to one another and our world."[16]

Mindless scrolling, like its near ancestor of mindless channel surfing, perpetuates mass trivialization. Our screens have a way of leveling pyramids of meaning such that all information, no matter how lofty and sacred, is reduced to rubble alongside ten million other digitized info-bricks. Entertainment value, novelty, clickworthiness, mass appeal, glitziness, and dopamine-hit potential become all important, rendering everything else trivial.

Missionally, the church goes wrong when it plays along. On a recent Super Bowl Sunday, the newsfeeds of millions lit up with cringey footage of a megachurch pastor kicking a Bible across the stage. It's become common for churches to host an "At the Movies" summer sermon series. Cardboard cutouts from the Marvel Universe line church foyers as pastors spend precious pulpit time expositing superhero movie plotlines with spurious connections to Scripture. Christians manufacture T-shirts of a Jesus silhouette

---

15 Rodney Clapp, "The Theology of Consumption and the Consumption of Theology," in *The Consuming Passion: Christianity and Consumer Culture*, ed. Rodney Clapp (Downers Grove, IL: InterVarsity, 1998), 188.

16 "Neil Postman Talk in LA."

sprawling through the air like Michael Jordan, slam-dunking the earth under the tagline "Air Jesus: The Ultimate High."

We must present Jesus as he is, which is anything but trivial. He is the serpent-crushing seed of the woman (Gen. 3:15). He is the Wonderful Counselor, Mighty God, Everlasting Father, and Prince of Peace (Isa. 9:6). He is the virgin-born (Matt. 1:20–23), storm-breaking (Mark 4:35–39), broken-body–mending (Luke 17:11–19), wave-walking (John 6:16–21), grave-defeating (20:1–29), reigning King (Eph. 2:6). He is the light of the world (John 8:12), the "I am" (8:58), the only door (10:7), the good shepherd (10:11), the resurrection and the life (11:25), and the vine (15:1). He is the image of the invisible God, by whom and for whom the universe was made and sustained (Col. 1:15–17). He is worshiped by angels (Heb. 1:6, 8). He is the slain Lamb of God (Rev. 5:6), the returning bridegroom (19:6–9), the King of kings and Lord of lords (19:16), the Alpha and Omega, the first and the last, the beginning and the end (22:13).

To minds conditioned by technology to trivialization, we present Jesus as the glorious Lord he is, not as another cheap gimmick.

Beyond Disinformation

To minds conditioned by disinformation—the loss of credible evidence—we present Jesus's death and resurrection as factual and historical, not fake news.

Another factor that renders us "adjusted to incoherence" is what Postman dubbed "disinformation." He defined it as "misplaced, irrelevant, fragmented or superficial information—information that creates the illusion of knowing something but which in fact leads one away from knowing."[17]

---

17 Postman, *Amusing Ourselves*, 107.

Imagine a social media user named Norm. On Thursday, Norm posts "Definitive Proof That the Apollo 11 Moon-Landing Was Filmed on a Hollywood Stage." On Friday, he reposts "University of Minnesota Study Finds That Chocolate Eclairs Contain Encomio Dioxine, a Proven Factor in Major Weight Loss."[18] On Saturday, he shares a link to a music store that is "Going Out of Business and Must Sell All Inventory by Midnight," including Fender Lonestar Stratocasters for "Just 50 Bucks Apiece!!!" Then comes Easter Sunday. Norm posts a graphic with the text "He Has Risen!" in an italic, somehow-more-unserious-than-Comic-Sans font over an animated GIF of a luminescent white Jesus glimmering outside the garden tomb.

Do you see how posting disinformation—a mistake just a finger-push away—can undermine someone's attempt to share the gospel's truth with credibility?

In first-century Athens, the public forum for the exchange of ideas was the Areopagus, where Paul gave his powerful missional plea recorded in Acts 17. Imagine if the apostle Paul opened with disinformation. "Did you guys hear Emperor Claudius's third wife, Valeria, was possibly born as a man?" Or "Did you know mixing goat milk with Mediterranean oyster juice can cure leprosy?" Then he'd follow up with, "Anyway, let me tell you now, as a reliable source, about the God you worship in ignorance, the one in whom we live and move and have our being, and who raised Jesus from the dead."

Paul wouldn't dare cast aspersions on the veracity of Christ's gospel by letting it come from the same tongue or quill as some

---

18 Neil Postman used this made-up example in an experiment he ran with colleagues and coworkers, an experiment he believed demonstrated how quickly people in the age of show business will accept obvious nonsense. See "Neil Postman Talk in LA."

wackadoodle conspiracy theorist or gullible spreader of baseless rumors. Paul warned Titus to silence empty talk and deception (1:10–11) and "avoid foolish controversies, genealogies, dissensions, and quarrels" (3:9). He exhorted Titus instead to "show integrity, dignity, and sound speech that cannot be condemned, so that an opponent may be put to shame, having nothing evil to say about [them]" (2:7–8). He warned Timothy about those with "an unhealthy craving for controversy and for quarrels about words, which produce envy, dissension, slander, evil suspicions, and constant friction" (1 Tim. 6:4–5), with an admonition to "avoid . . . irreverent babble" (1 Tim. 6:20; cf. 2 Tim. 2:16; Heb. 13:9).

Given our truth-telling mission, Christians should be the most cautious about posting and reposting unsubstantiated, sensationalistic content and conspiracy theories.[19] This becomes especially vital as AI technology increasingly blurs the line between facts and fakery, the genuine and the generated, and the human and pseudo human. The presence of Christians on social media should add to the net amount of truth, goodness, and beauty while reducing the amount of nonsense in cyberspace.

Beyond Disintegration

To minds conditioned by disintegration—the loss of a grand unifying story—we show Jesus as the supreme integration point in whom everything finds meaning, not as a free-floating data point.

Grand stories about reality, Postman argued, help fortify our minds against absurdity, chaos, and incoherence. He showed the superiority

---

19 Paul's admonition to the Philippians provides guidance here. "Whatever is true, whatever is honorable, whatever is just, whatever is pure, whatever is lovely, whatever is commendable, if there is any excellence, if there is anything worthy of praise, think about these things" (Phil. 4:8). Note well, Paul begins his list with "Whatever is true."

of the age of faith over the age of show business: "In the belief system of the middle ages there existed an ordered comprehensible worldview beginning with the idea that all knowledge and goodness come from God. . . . The situation we're in presently is much different. . . . There is no consistent integrated conception of the world which serves as the foundation on which our edifice of belief rests."[20]

What we need, according to Postman, is some "god" (with a lowercase *g*)—some supreme integration point.[21] For those who have problems with the g-word, Postman suggested "narrative" as an alternative: a grand story of where we came from, who we are, and where we are headed. He weighed several contemporary options and found them wanting.

There is "the god of economic utility," in which the uttermost goal of education is "to teach the young how to make a living not how to make a life." There is the related "god of consumership," catechizing students to believe "they are not what they do but what they own."[22] There is "the great god of technology," to help the young become what technology will make them become. Anticipating the infusion of diversity and inclusion dogmas and critical race theory into education,[23] corporations, media, and entertainment

---

20 "Neil Postman: The End of Education," directed by Robert Dinozzo, Into the Classroom Media (New York Library Center for the Humanities, 1996), YouTube. This lecture is largely a live reproduction of content from Postman's *The End of Education* (Vintage, 1996).

21 "End of Education."

22 Hence the observation of Duke University economist Thomas Naylor about students in the 1990s asked to write personal strategic plans: "With few exceptions, what they wanted fell into three categories: money, power, and things—very big things including vacation homes, expensive foreign automobiles, yachts, and even airplanes." What they wanted from faculty, Naylor observed, was simply "Teach me how to be a money-making machine." Naylor, "Redefining Corporate Motivation, Swedish Style," *The Christian Century* 107 (1990): 566–70.

23 In education, this is what Jonathan Haidt dubs "Social Justice University" in contrast to "Truth University." See Greg Lukianoff and Jonathan Haidt, *The Coddling of the American Mind: How Good Intentions and Bad Ideas Are Setting Up a Generation for Failure* (New York:

today, Postman critiqued "the god of tribalism and separatism," worshiped then under the guise of "multiculturalism."[24]

In a culture devoid of a grand meaningful story and turning to inadequate, ultimately dehumanizing stories, it is the Christian's task to tell the better story.[25] Tell the best story. Tell the Bible's story of creation, fall, redemption, and consummation. Tell in the pulpits the good news of Jesus's perfect life, substitutionary death, and bodily resurrection. Rehearse that story in the sacraments. Catechize your children into that story. Heavily stock a church library with the great theological minds who exposit that story.[26] Embody that story by living a life marked by grace and love in the context of real flesh-and-blood, first-name-basis communities.[27]

Penguin, 2018). See also my work *Confronting Injustice without Compromising Truth: 12 Questions Christians Should Ask About Social Justice* (Grand Rapids, MI: Zondervan, 2020).

24 "As a point of view that stresses above all else love of tribe implying separateness from if not hostility to others, then multiculturalism is indeed a dangerous god to serve. . . . The theme of schooling would then be divisiveness, not sameness." Postman, "End of Education." The massive body of empirical evidence that diversity, equity, and inclusion training decreases diversity in organizations, lowers morale, reinforces biases, and negatively affects the mental and physical health of minorities supports Postman's warnings. See Thaddeus Williams, "DEI's Grape-Nuts Problem," *World*, March 14, 2024, https://wng.org/.

25 J. R. R. Tolkien adds, "The Gospels contain . . . a story of a larger kind which embraces all the essence of fairy-stories. . . . Because this story is supreme; and it is true. Art has been verified." Tolkien, "On Fairy Stories," in *Essays Presented to Charles Williams* (London: Oxford University Press, 1947), 83–84. As Tolkien's friend C. S. Lewis came to believe, largely through Tolkien's help, "the story of Christ is simply a true myth." Lewis, "Letter to Arthur Greeves, October 1, 1931," in *Collected Letters, Vol. 1: Family Letters 1905–1931*, ed. Walter Hooper (San Francisco: Harper San Francisco, 2004), 976–77.

26 Why shouldn't an average church attendee show up on Sunday to return a copy of Calvin's *Institutes* and head home with Chesterton's *Orthodoxy* (and maybe a volume from Tolkien's Middle-earth or Lewis's Narnia for the kids' bedtime)? If your church doesn't have any such library, start one.

27 The local embodied church becomes essential in our age because of the epidemic of loneliness that scrolling technologies unleash. "With personal computers the average person can . . . vote at home, shop at home, get all the information they wish at home, and thus make community life entirely unnecessary." Postman, "Neil Postman Talk in LA."

Furthermore, we must clearly communicate our Christian world-view *as a worldview*, a Christ-centered vision that makes sense of *all* reality. Christianity scratches humanity's deepest existential itches for relationship, freedom, mystery, beauty, awe, hope, and more. It is a coherent system of truth robust enough to inspire the Aquinases and Alvin Plantingas of philosophy, the Dostoyevskys and Marilynne Robinsons of literature, the Blackstones and Wilber-forces of statecraft, the Frederick Douglasses and Lee Jong-raks of human rights, the Francis Bacons and George Washington Carvers of science, the Bachs and Handels of music, the Van Goghs and Rembrandts of painting, the Bonhoeffers and Sophie Scholls of fighting tyranny, the Lewises and Tolkiens of creating fantasy, the Pascals and John Lennoxes of mathematics, and more.

One of many factors that gives Christians an advantage in the ages of show business and scrolling is what Abraham Kuyper recognized as the shared human need for a "life-system" or "a unity of view" because "the question about the origin, interconnection and destiny of everything that exists cannot be suppressed."[28] Yet, as Postman notes, "The computer and its information cannot answer any of the fundamental questions we need to address to make our lives more meaningful and more humane. It is a magnificent toy that distracts us from facing what we most need to confront: spiritual emptiness."[29]

To those disoriented by Postman's "world of fragments, where events stand alone, stripped of any connection,"[30] we must display Christianity's beautiful, radically humanizing, and God-glorifying coherence.

---

28 Abraham Kuyper, *Lectures on Calvinism* (Grand Rapids, MI: Eerdmans, 1999), 113. Kuyper adds, "And the *veni, vidi, vici* [I came, I saw, I conquered], wherewith the theory of evolution with full speed occupied the ground in all circles, inimical to the Word of God, and especially among our naturalists, is a convincing proof how much we need unity of view."

29 "Neil Postman Talk in LA."

30 Postman, *Amusing Ourselves*, 110.

May we carry forward the torch of Kuyper's famous credo: "There is not a square inch in the whole domain of our human existence over which Christ, who is Sovereign over all, does not cry: 'Mine!' "[31] The Jesus who alone can save us from wrath and sin is the same Jesus who can save us from the incoherence of our age of scrolling.

Discussion Questions

1. To Williams, the negative effects of scrolling actually enhance the gospel's beauty. He writes there are many factors that give Christianity an advantage in the age of scrolling, listing the need for a "coherent" worldview as one of them. What are some other specific advantages of the gospel in our scrolling age?

2. Why do you think Christians today feel the need to apologize for sharing the gospel in our culture? When have you felt this temptation yourself? What do you find are the hardest aspects of Christianity for our culture to receive?

3. Postman argued that the age of show business is both a technology and an epistemology. What other technologies have transformed the way humans think and understand the world? What might be the next kind of technology that will utterly reshape human understanding?

4. What is the rub between the "one way" of Jesus and the infinite possibilities of a digitalized age? Does a scrolling society have a framework for the concept of a singular, exclusive truth?

---

31 Abraham Kuyper, inaugural lecture at the Free University of Amsterdam, October 20, 1880, quoted in *Abraham Kuyper: A Centennial Reader*, ed. James D. Bratt (Grand Rapids, MI: Eerdmans, 1998), 488.

9

"Unfit to Remember"

The Theological Crisis of Digital-Age Memory Loss

Nathan A. Finn

DEMENTIA HAS CAST A CRUEL SHADOW over my family for many years. Both my mother and my wife's mother battled dementia for more than a decade. Mom died in September 2018. We lost my mother-in-law fifteen months later, in December 2019. Both women were diagnosed with forms of dementia when they were in their fifties. Both were younger than most dementia patients. Both suffered for years. So did their families and closest friends as they navigated the effects of a terminal brain condition on a loved one. In the end, as their brains quit communicating with the rest of their organs, we lost these women far earlier than seems right.

Dementia is a thief. It steals your life by impairing your memory. Every dementia patient's story is different. Mom would remember things incorrectly. In the same conversation, she would recall some details perfectly, but others would seemingly come out of left field, leading to confusion and sometimes conflict. Her personality changed in troubling ways. Then she declined rapidly, over about six months. In many ways, my mother-in-law's situation was even worse: she simply stopped remembering. For the final years of her life, she didn't know any of us. She had forgotten everything. It was a long goodbye before she finally passed away.

The church is always tempted by a form of spiritual dementia. But unlike the medical condition, this type of dementia is self-selected, whether passively through neglect or actively through rebellion. In the fourth century, the Arians forgot who Jesus really was and led much of the Eastern church into heresy. In the medieval West, various popes and other ecclesiastical leaders forgot what it meant to be shepherds and embraced moral profligacy while courting political power. Christians from the 1500s to mid-1800s forgot what it meant that all people are created in God's image as they enslaved Africans and their descendants for centuries. In the twentieth century, many German Protestants forgot what it meant to obey God rather than men and supported the murderous Nazi regime of Adolf Hitler. In our day, many professing believers have forgotten what it means that God created humans male and female, or that his intention for men and women is a lifelong one-flesh union, or that humans are embodied creatures, or that humans have greater inherent dignity than other creatures, or that every human possesses dignity.

Spiritual dementia is incompatible with Christian faithfulness. As both a church historian and a pastor, I'm increasingly convinced that life in the digital age compounds the potential for losing our theological and ethical memory. Our highly digital existence enables us to remain connected constantly through what Chris Martin calls the "social internet."[1] While this technology is convenient in many ways, it often drives us to narrowly prioritize the present moment at the expense of the bigger stories and wider contexts that make sense of our lives. Such stories certainly include the narrative arc of Scripture, which has rightly been called "the true story of the whole world."[2] But it also includes lesser, though still important, stories that form us: The stories of our families. The stories of our communities. The stories of our nations. The stories of our vocations. And the story of Christianity—church history.

Arrogance of Amnesia

No technology is simply a neutral tool to be used for whatever utilitarian purpose we deem appropriate.[3] This is certainly true of the internet, the influence of which touches almost every part of our lives. Like Sauron's malevolent ring from J. R. R. Tolkien's *The Lord of the Rings*, the internet seeks to subdue every rival and rule them all. It demands our full attention. It seduces us into increased allegiance. And prolonged, unreflective exposure to the internet withers our souls until we become diminished shades who forget what life was once like in the real world. The internet conditions us to embrace what I once heard church historian Timothy George refer to in a different context as the "arrogance of amnesia." We no longer remember who we are and how we got here, but we assume

---

1 Chris Martin, *Terms of Service: The Real Cost of Social Media* (Brentwood, TN: B&H, 2022), 6–7.

2 Michael W. Goheen and Craig G. Bartholomew, *The True Story of the Whole World: Finding Your Place in the Biblical Drama*, rev. ed. (Grand Rapids, MI: Baker, 2020).

3 See Jason Thacker, "Simply a Tool? Toward a Christian Philosophy of Technology and Vision for Navigating the Digital Public Square," in Jason Thacker, ed., *The Digital Public Square: Christian Ethics in a Technological Society* (Brentwood, TN: B&H Academic, 2023), 3–26.

we know more than everyone who came before us because of our constant access to nearly limitless content.

None of this digitally induced amnesia would surprise Neil Postman. Though he died shortly before the digital revolution took off, his writings about the relationship between television and memory proved prescient. In *Amusing Ourselves to Death*, Postman argued "television is a speed-of-light medium, a present-centered medium. Its grammar, so to say, permits no access to the past."[4] I'd suggest this is far more the case with more recent forms of digital media—particularly smartphones—which have the potential to keep us mediated nearly every moment, no matter where we are. Even more than TV, the internet keeps viewers hooked on the present, their eyes glued to the momentary distractions passing across their screens or feeds.

Postman argued that the coming technological dystopia looked more like Aldous Huxley's *Brave New World* (1932) of pleasure-induced compliance than like the totalitarian coercion of George Orwell's *1984* (1949). Generally, I agree with Postman on this point. But in at least one important way, our digital age calls back to Orwell's famous novel. In that story, the ominously misnamed Ministry of Truth manipulates the population by controlling all information. Whenever new propaganda is published, all older material is deposited in chutes found throughout the building that empty into a giant incinerator hidden away in the building's recesses. These chutes are called memory holes. The past is regularly pushed down the memory hole, thus keeping the populace focused on the ever-reinvented present.[5] In modern usage, the term

---

4 Neil Postman, *Amusing Ourselves to Death*, 20th anniversary ed. (1985; repr., New York: Penguin Books, 2005), 136.

5 George Orwell, *1984*, 75th anniversary ed. (1949; repr., New York: Signet Classic, 2023), 37–38.

"memory hole" has come to refer to any attempt to revise the past by ignoring it, suppressing it, or simply rewriting it.

The internet, and especially the social internet, functions like an ever-present, ever-growing memory hole. I don't simply mean people are always scrubbing or reframing "inconvenient" information from the internet and reinventing themselves, their organizations, or their causes—though that is certainly true. Rather, I mean that the internet is *designed* to keep us fixated on the present, rewarding and reinforcing engagement with what is current, new, and seemingly urgent.

In his much-discussed book *The Shallows*, Nicholas Carr argues, "Our use of the Internet involves many paradoxes, but the one that promises to have the greatest long-term influence over how we think is this one: the Net seizes our attention only to scatter it."[6] This isn't a bug but a feature of the internet. Carr's book shows how our brains are rewired by nearly every aspect of the internet, how companies exploit this, and why we should be concerned. One cognitive casualty is our personal memory, which can't help but atrophy the more we depend on the artificial digital memories we access online. Carr's assessment is pointed: "The Web is a technology of forgetfulness."[7]

While the internet's erosion of our ability to remember is troubling, this is coupled with the reality that the digital world almost compels us to be focused on the present. The twenty-four-hour news cycle that accompanied the rise of cable news networks in the late twentieth century seems quaint compared to the sheer number of options (both good and bad) and the cacophony of competing voices available online. Much of the social internet is driven by

---

6 Nicholas Carr, *The Shallows: What the Internet Is Doing to Our Brains*, updated ed. (New York: Norton, 2020), 118.

7 Carr, *The Shallows*, 193.

emails addressing the needs of the moment, group texts that keep us engaged in ongoing conversation, social media updates that keep us plugged in to what's trending, and influencers who promote contemporary trends. Felicia Wu Song rightly refers to our digital lives as "mobile, social, and infinitely novel."[8]

A generation ago, Postman compared television to a mirror. The technology was designed to reflect the ever-changing preferences of the moment rather than point us to the wisdom of the past. He quoted broadcast journalist Bill Moyers, who referred to the 1980s as an "anxious age of amnesiacs" and warned that "Americans seem to know everything about the last twenty-four hours but very little of the last sixty centuries or the last sixty years."[9] Television dulled historical memory, and while some programming certainly focused on historical topics, these periodic forays into the past were hardly an antidote to the present-driven demands of the medium. Postman lamented, "We do not refuse to remember; neither do we find it exactly useless to remember. Rather, we are being rendered unfit to remember."[10]

In *Technopoly*, Postman advised that "when we admit a new technology to the culture, we must do so with our eyes wide open."[11] Few heeded these sorts of warnings when the digital age began, nor did many of us have concerns when the social internet first became part of the fabric of our lives. We have reaped the consequences with our diminished ability to remember the past, which should be especially troubling for believers who understand we are members

---

8   Felicia Wu Song, *Restless Devices: Recovering Personhood, Presence, and Place in the Digital Age* (Downers Grove, IL: IVP Academic, 2021), 161.

9   Bill Moyers, speech (National Jewish Archive of Broadcasting, New York, March 7, 1984), cited in Postman, *Amusing Ourselves*, 137.

10 Postman, *Amusing Ourselves*, 137.

11 Neil Postman, *Technopoly: The Surrender of Culture to Technology* (New York: Vintage, 1993), 7.

of the communion of saints that stretches back to creation and extends into eternity. God desires for us a better way.

Remembering and Recounting

In her insightful book *The Spiritual Practice of Remembering*, Margaret Bendroth argues that in the Christian tradition, "remembering and forgetting were decisions with moral consequences."[12] As a historian and archivist, Bendroth is concerned that believers (and others) remember rightly the Christian past—the good and the bad—so we might better embody present faithfulness. In the Pentateuch, God remembered his covenants as a sign of his unchanging character (Gen. 9:15–16; Ex. 2:24; 6:5; Lev. 26:42, 45). God also commanded his people to remember both his mighty acts (especially the deliverance of Israel from Egyptian enslavement) and his commandments (Ex. 3:15; 13:3; Num. 15:39–40; Deut. 5:15; 7:18; 16:12). These themes continued throughout the Old Testament, especially the Psalms and the Prophets. As the psalmist exclaimed,

> I will remember the deeds of the LORD;
> yes, I will remember your wonders of old. (Ps. 77:11)

Importantly, this spiritual remembering was not just a facet of individual devotion but a multigenerational practice. God's commandments were to be told and retold as a formative reminder of who—and whose—Israel was:

> Hear, O Israel: The LORD our God, the LORD is one. You shall love the LORD your God with all your heart and with all your

12 Margaret Bendroth, *The Spiritual Practice of Remembering* (Grand Rapids, MI: Eerdmans, 2013), 128.

soul and with all your might. And these words that I command you today shall be on your heart. You shall teach them diligently to your children, and shall talk of them when you sit in your house, and when you walk by the way, and when you lie down, and when you rise. You shall bind them as a sign on your hand, and they shall be as frontlets between your eyes. You shall write them on the doorposts of your house and on your gates. (Deut. 6:4–9)

Remembrance was at the heart of Israel's devotion and shaped the rhythm of the people's spiritual lives, both individually and corporately.

This emphasis carried over into the early church. Notably, Jesus cited Deuteronomy 6:5 as the first and greatest commandment, which the Holy Spirit led Matthew to record for the earliest believers (Matt. 22:36–38). Paul referred to traditions he handed down to the early church (1 Cor. 11:2; 2 Thess. 2:15; 3:6). Peter reminded believers of truths they already knew (2 Pet. 1:12–13). Jude 3 commanded Christians "to contend for the faith that was once for all delivered to the saints." These texts imply acts of spiritual remembering that were consistent with, and an expansion of, the earlier commands for Israel to remember and pass on the truths of God and his ways.

Postman argued that "history is of value only to someone who takes seriously the notion that there are patterns in the past which may provide the present with nourishing traditions."[13] As believers, we certainly believe this is the case. God calls us to remember the past. This includes first and foremost Scripture, which recounts

---

13 Postman, *Amusing Ourselves*, 136.

the words and ways of God from creation to consummation and forms our biblical worldview. But it also applies to the Christian past, from the end of the New Testament age to yesterday. The digital age has made our memories sluggish. We must retrain ourselves to be what the eighteenth-century Anglican evangelical Ambrose Serle called "Christian remembrancers," reminding ourselves regularly of biblical truths and their implications for our lives.[14]

Cultivating Christian Remembering

England declared war against Germany on September 3, 1939, thereby entering a second European war in less than a generation. A little over a month later, C. S. Lewis delivered a sermon in the Church of St Mary the Virgin in Oxford, titled "Learning in War-Time." Lewis told the congregation of students that there is no such thing as a normal life, so they should continue to pursue their studies just as much in a time of war as in a time of peace. He also spoke eloquently about the need to understand the past:

> Most of all, perhaps, we need intimate knowledge of the past. Not that the past has any magic about it, but because we cannot study the future, and yet need something to set against the present, to remind us that the basic assumptions have been quite different in different periods and that much which seems certain to the uneducated is merely temporary fashion. A man who has lived in many places is not likely to be deceived by the local errors of his native village: the scholar has lived in many times and is therefore in some degree immune from the great cataract

14 Ambrose Serle, *The Christian Remembrancer: Short Reflections on the Faith, Life, and Conduct of a Real Christian* (1799; repr., Harpenden, UK: Gospel Standard Trust, 2017).

of nonsense that pours from the press and the microphone of his own age.[15]

The digital age is certainly characterized by "the great cataract of nonsense." We need to resist the tyranny of the present, which is the warp and woof of the internet, by cultivating habits that help us remember the stories that matter most. We need to find ways to practice the spiritual discipline of Christian remembering. Here are a few suggestions for how you might do that:

- For your regular devotional time, read from a physical Bible and incorporate creeds, confessions, liturgies, and older hymns into your prayer life.[16]
- Supplement your Scripture reading with edifying biographies of faithful Christian leaders.
- Consider following C. S. Lewis's suggestion to read one old book for every new book—or, "if that is too much for you, you should at least read one old one to every three new ones."[17]
- Listen to podcasts or audiobooks focusing on church history during your daily commute, while exercising, or when doing chores around your home.
- Lean into the historical continuity of Christianity. At church when you hear Scripture preached, take the Lord's

15 C. S. Lewis, "Learning in War-Time," in *The Weight of Glory: And Other Addresses* (1941; repr., New York: HarperOne, 2001), 58–59.

16 Jonathan Gibson has compiled several excellent devotional resources that include liturgies that draw on historical sources. See *Be Thou My Vision: A Liturgy for Daily Worship* (Wheaton, IL: Crossway, 2021); *O Come, O Come, Emmanuel: A Liturgy for Daily Worship from Advent to Epiphany* (Wheaton, IL: Crossway, 2023); *O Sacred Head, Now Wounded: A Liturgy for Daily Worship from Pascha to Pentecost* (Wheaton, IL: Crossway, 2024).

17 C. S. Lewis, "On the Reading of Old Books," in *God in the Dock: Essays on Theology and Ethics* (Grand Rapids, MI: Eerdmans, 1970), 201.

Supper, watch someone get baptized, greet one another, take up tithes and offerings, and engage in other worship practices, self-consciously remind yourself that believers have been doing these things for two millennia—and are doing them still today, all over the world. Your story is part of a much larger story, and your pursuit of godliness can be shaped by the example of saints from bygone days.

For pastors or church leaders, I encourage you to include Christian history in your church's discipleship menu:

- When you preach and teach, cite regularly from past Christian leaders who offer evergreen spiritual wisdom. (I know Spurgeon is great, but try to stretch yourself.)
- Incorporate creeds or selections from your tradition's confessional statements into your church's liturgy.
- Consider offering a short course for church members on church history or the history of your particular ecclesiastical tradition.
- Host a reading group that works through a short survey of church history or the biography of a famous saint.

These are just a few ideas, but this is the larger point: you must be intentional about making the practice of Christian remembering a part of your strategy for spiritual formation. In the forgetful age of the social internet, *remembering is resistance*. So, as you seek ways to rightly order your digital life, make sure you include counterformative practices that remind you that memory matters for our spiritual flourishing. After all, the present always lasts only a moment, but the wisdom of the Christian past will continue to speak into eternity.

Discussion Questions

1. If we agree with Finn that the internet is designed to shackle us to the present, how will this default setting negatively affect Christian discipleship?

2. "Americans seem to know everything about the last twenty-four hours but very little of the last sixty centuries or the last sixty years." Discuss Bill Moyers's quote. In your experience, is it a generally accurate representation? Do you find this lack of historical knowledge has accelerated with younger generations?

3. Look through Finn's list of practices for our "memories." Which one sticks out the most? What other memory-training disciplines might you incorporate into your life to resist the presentist focus of a scrolling world?

4. C. S. Lewis wrote "Learning in War-Time" to combat the lie that says we can't spend time on seemingly frivolous things (like learning) when so many important things are happening. How do digital marketers leverage this same lie against us now? How might wisdom refute this lie?

PART 3

HOW THE CHURCH CAN BE LIFE IN A "SCROLLING TO DEATH" WORLD

10

Use New Media Creatively but Cautiously

Video as Case Study

G. Shane Morris

AT THE HEART OF *AMUSING OURSELVES* is a simple argument: that certain ideas or assumptions about the world are embedded in our tools of communication, that television as a tool is unlike print in that it delivers entertainment rather than information, and that this tendency reshapes our minds and society in troubling ways.[1] If we had to look back across the forty years since Postman made his argument and ask what media or technology development has most changed the landscape, our top candidate might be online video.

---

1 Neil Postman, *Amusing Ourselves to Death*, 20th anniversary ed. (1985; repr., New York: Penguin Books, 2005), 78.

Anyone who doubts that internet-based video has edged out not only books but even television need only talk to a young person. I once saw a meme poking fun at how thoroughly video sharing apps now dominate our discourse and thought. According to the meme, when people say, "I read somewhere . . ." "I learned in a documentary . . ." or "Someone was telling me . . ." what they really mean is "I saw it on TikTok."

We might say the same for older platforms like YouTube and competing social media scrolling video platforms like Reels on Instagram and Facebook. Postman noticed that television in his time was the master medium in control of all others and that it was inherently focused on show business to the extent it trained viewers to expect everything to *be a show*. That's now largely true of social media, specifically video.[2] Yet internet video has, in important ways, taken us beyond mere entertainment.

Short, Disjointed, Individualistic

Let's start with the most obvious difference: while TVs and the screens on which we view internet videos bear a superficial resemblance, the internet and social media algorithms have made video content much shorter, much more disjointed, and much more individualistic.

Part of Postman's criticism of television was that it prized bite-size clips and "constant stimulation through variety, novelty, action, and movement."[3] His examples of this staging, like the *MacNeil/Lehrer NewsHour*, seem sedate and in-depth by today's standards. An hour-long program on one subject? The most successful TikTok videos last between twenty-one and thirty-four seconds![4]

---

2 Postman, *Amusing Ourselves*, 78.

3 Postman, *Amusing Ourselves*, 105.

4 Chris Stokel-Walker, "TikTok Wants Longer Videos—Whether You Like It or Not," *Wired*, February 21, 2022, https://www.wired.com/.

The very name of today's most popular video sharing app alludes to the sound of a clock, an invention Postman said forever changed human life by dicing up our days into mechanically mediated units and disconnecting us from eternity.[5] He argued that only so much thinking could be depicted visually on-screen because complexity, nuance, and qualifications are excluded by the nature of the medium. Indeed, "visual stimulation" often substitutes entirely for thought.[6] Experts tell us the average person takes ten to fifteen seconds simply to process a new piece of information.[7] It takes far longer to consider it and formulate a response. By that time, a new video is queued up and tailored teasers beckon us: "Watch this next!" Such insectile intervals mean human cognition is literally incapable of doing anything with information presented in this format. We experience it more with our reflexes than with our minds.

Consider also that each fleeting clip is largely irrelevant to the clip that preceded it and is algorithmically curated for each user. When Postman critiqued the "Now . . . this" rhythm that pervaded television and trivialized the world, newscasters and showrunners were still consciously making decisions about what came on next.[8] These days, little or no human agency is involved with the flow of content we consume. Instead, a computer decides what the user will see at the next flick of a finger based on what it calculates will keep that user scrolling. What you see on your phone is not what I see on mine. Our respective feeds are optimized to suck us in *as individuals*.

5   Postman, *Amusing Ourselves*, 11.

6   Postman, *Amusing Ourselves*, 105.

7   "Processing Time," Salt & Light: The Speech Therapy Company, December 7, 2018, https://www.speechtherapycompany.co.uk/.

8   Postman, *Amusing Ourselves*, 99, 104–5.

Everyone a Performer

Internet video has universalized the screen's tendency to turn everyone who appears on it into an entertainer. Postman wrote that on television, the "impression of sincerity, authenticity, vulnerability or attractiveness" substitutes for genuine credibility.[9] But at least people who appear on television have to submit a résumé. Anyone can record and post content to video-sharing apps and potentially rack up millions of views. And, unlike on television, the typical "creator" or "influencer" online is her own cast, camera crew, producer, fact-checker, and publicist, resulting on average in lower quality and more superficial content that relies on gimmicks over substance.

Even older and comparatively longer-form video platforms like YouTube are a disorienting parade of users pulling cartoonish faces, wearing minimal clothing, flashing bright colors, or displaying clickbait captions simply to stand out from the crowd. Videos longer than ten minutes often feature a TikTok-length "teaser" for an upcoming highlight. Given these incentives and limitations, content creators and consumers rarely distinguish between what is true and will inform, instruct, or edify versus what will get clicks, views, and shares.

From Entertainment to Distraction

All this means that video on the internet has largely transcended Postman's category of entertainment and aims instead at something more primal, trivial, and chemical: *distraction.*

An hour on TikTok trains our brains and bodies to expect constant stimulation and novelty—something Postman's preferred medium of the printed word simply cannot deliver and which even television often subordinates to production value. The results of this

---

9   Postman, *Amusing Ourselves,* 102.

training live in our nervous systems. In his 2016 *New York Magazine* essay, "I Used to Be a Human Being," Andrew Sullivan confesses what happened when he tried reading books after extended time on social media: "That skill . . . began to elude me. After a couple of pages, my fingers twitched for a keyboard."[10]

Something has changed since Postman wrote, and we can all feel it. Entertainment has been dethroned by a more compulsive force. Who among us has not sat down to watch a movie only to find our attention drifting to a smaller screen in our pocket? As a society, we've replaced the demand for entertainment with an appetite for *distraction*, and nowhere do we see this trend more clearly than in online videos.

Some Mediums Are Better

Postman's pessimism about the entertainment culture of his time makes it difficult to mine his work for insights or solutions to today's video-based dopamine culture. He treats television as a medium with little to recommend it other than its ability to amuse, even claiming television is at its most dangerous when its aspirations are *high*—when it seeks to educate, inform, or spiritually edify.[11]

He was not an "epistemological relativist." He genuinely viewed some ways of "truth-telling" as superior to others and therefore as healthier influences on cultures that adopt them.[12] This is why he argued a print-based culture and the "typographic minds" it nurtured were preferable to a televised culture. Yet he cautioned against assuming we know everything and urged readers to keep an open mind about "what benefits may come from other directions."[13]

---

10 Andrew Sullivan, "I Used to Be a Human Being," *New York Magazine*, September 2016, https://nymag.com/.

11 Postman, *Amusing Ourselves*, 16.

12 Postman, *Amusing Ourselves*, 24.

13 Postman, *Amusing Ourselves*, 29.

We can learn something from this insistence that some mediums of communication are better than others. This implies human beings are "wired" to learn and think in some optimal way and that oral, print, broadcast, and internet technologies are better or worse at approximating that ideal. And if we can get worse at communication over time, why can't we get better?

Postman's suspicion toward imagery is especially important to wrestle with as we consider solutions to our video distraction problem. He claimed that Moses's command against graven images not only forbids depicting Yahweh but implies a broader prohibition on art or visual communication.[14] This conclusion is difficult to defend in light of God's later command that Moses place golden cherubim on the lid of the ark (Ex. 25:18) or the vibrant imagery incorporated in the Jewish temple (1 Kings 7:40–44; 2 Chron. 3:11). Nevertheless, Postman's point that people are trained in "abstract, universal" thinking by attending to the written Scriptures is well taken.[15]

We were made *by* the Word *for* the word (John 1:1–3; 2 Tim. 3:16–17). There is a nonnegotiability to the medium through which God has chosen to reveal himself to mankind. Writing is special. This view resonates with evangelical Protestants, holding as we do to the Reformation principle of *sola Scriptura*. Postman acknowledged this fact, crediting Protestant love of the Bible for the historically high literacy in colonial America.[16]

Our Desires Point Somewhere

Yet our eyes were made for more than reading. We have color vision, extraordinarily subtle and acute depth perception, and brains

---

14 Postman, *Amusing Ourselves*, 9.
15 Postman, *Amusing Ourselves*, 9.
16 Postman, *Amusing Ourselves*, 32.

disproportionately dedicated to processing what we see. This is the very reason television and internet video, as mediums, are so appealing. We enjoy watching them because they show us the world, and we were made to enjoy watching the world![17] There are several conclusions we can draw from this fact.

First, our desires point somewhere. The reason internet pornography, for instance, has become addictive to millions is that humans were designed to enjoy the sight of a naked member of the opposite sex, specifically a spouse. "Men feel sexual desire," wrote C. S. Lewis, because "there is such a thing as sex."[18] In other words, the desire for sex and sexual *sights* was originally good and points to a real fulfillment. In online pornography, that desire is twisted and distorted into something rapacious, isolating, and dehumanizing—often by means of video. Yet no one thinks pornography proves that sexual desire itself is bad.

In much the same way, online video, even the silly or trivial sort, appeals to us powerfully because we were made to *see*, and to *see in community*.

Think of the strange phenomenon of "stitched" TikTok videos or Instagram reels that split an already tiny screen between the focal

---

17 See, for example, what Augustine said on the allure of color and its power to draw his attention, even away from contemplating God: "A delight to my eyes are beautiful and varied forms, glowing and pleasant colours. May these get no hold upon my soul; may God hold it! 'He had made these sights and they are very good' (Gen. 1:31). But he is my good, not these. They touch me, wide awake, throughout the day, nor do they give me a moment's respite, in the way the voices of singers, sometimes the entire choir, keep silence. The very queen of colours, which bathes with light all that we see, wherever I may be during the day, comes down upon me with gentle subtlety through many media, while I am doing something else and not noticing it. But the light makes its way with such power that, if suddenly it is withdrawn, it is sought for with longing. And if it is long absent, that has a depressing effect on the mind." Augustine, *Confessions*, trans. Henry Chadwick, Oxford World's Classics (New York: Oxford University Press, 1992), 209.

18 C. S. Lewis, *Mere Christianity* (1952; repr., New York: Harper Collins, 2001), 136.

video (someone doing a stunt, talking, or visiting a destination) and a second video showing a user reacting to the first video. What is the point of this odd ritual if not to tickle (and often exploit) the human desire to see *in community*—to have a friend beside you sharing your amazement or laughter or shock? The reason such videos trigger a dopamine release is that they appeal to instincts and purposes in us which predate videos, smartphones, or even television and writing. In an important sense, they are mimicking real human life and profiting from the resemblance, however imperfect it may be.

Limits Can Be Good

Yet the unsettling impression left by a person watching a video of a person watching a video points to a second conclusion: the limits our bodies impose on our time, our relationships, our work, and our worship can be good. These limits are not always obstacles we should strive to overcome with technology. In this sense, Postman's warnings ring clearer than ever.

Try as we might, we can't conduct our friendships exclusively over screens. Physical presence and therefore spatial limitations are required at some point. The idea of living out a marriage totally online strikes us as absurd. Bodily presence is a nonnegotiable in such a relationship, and periods when couples resort to the telephone or video chats are always marked by expressions of eagerness to be back together in person.

Or consider the sacred. Postman warned of the American belief, shared by many preachers, that every form of discourse can be televised.[19] He cited a plea by Pat Robertson for churches to get involved with television, because in Roberton's words, "the needs are

19 Postman, *Amusing Ourselves*, 117–18.

the same, the message is the same, but the delivery can change."[20] Postman called this gross naivete, and it's not hard to see why.

Christianity is a physical religion involving tangible sacraments, face-to-face fellowship, and the participation in worship and word in physical spaces. Converts are baptized in real water, communed with real bread and wine, and welcomed into a real, local body of believers. It has always been so, and to demand Christianity renegotiate these inheritances is to demand it become a fundamentally different, less embodied religion.

And, as Postman warned, screens also have a strong secularizing influence, saturated as they are with "memories of profane events" and deep associations with the "commercial and entertainment."[21] Our bodies limit us when it comes to honoring God, and that was deliberate on God's part. We should not seek to overcome his design—here or in the many other areas where new media can never replace real life.

Practical Applications

In light of our discussion of how internet video can hijack or corrupt good features of our human nature, we need to develop a strategy to help us move forward wisely in today's flashy world of distraction. We begin by mounting a strong defense through cultivating self-awareness and healthy desires, then gradually go on the offensive by harnessing video's allure for godly and humanizing purposes. Here's what that might look like, step by step.

1. Realize What's Happening to You

In an especially prescient passage in *The Screwtape Letters*, C. S. Lewis's senior tempter instructs his nephew on how to get a human to waste

20 Postman, *Amusing Ourselves*, 117–18.
21 Postman, *Amusing Ourselves*, 119.

his life doing "Nothing": "You no longer need a good book, which he really likes, to keep him from his prayers or his work or his sleep," writes the old devil. "You can keep him up late at night," doing things as aimless as "staring at a dead fire in a cold room[,]" his mind flickering "over it knows not what and knows not why, in the gratification of curiosities so feeble that the man is only half aware of them."[22]

This, explains Screwtape, is the power of Nothing. This is how you destroy a human with distraction.

Few warnings are more applicable to the negative effects of internet video in our time, especially the kind of video that lacks even the entertainment value Postman critiqued. When you get right down to it, scrolling isn't that fun. It's just something to do—something that triggers enough of a dopamine release to keep us coming back time and again.

That's why the first practical application of our case study is this: when we're mindlessly flicking through short videos, we should simply stop and realize what's happening to us. In a rare moment of optimism, Postman allowed that "no medium is excessively dangerous if its users understand what its dangers are."[23] This is just as true in our time as it was in the 1980s. Watching a television while *knowing* that it is restaging the world makes the television significantly less harmful. In the same way, being conscious of the dangers of new media, especially the kind that appeals to our visual and social appetites, is a crucial step in avoiding those dangers.

2. Ask Good Questions

The next step is to give up trying to fight nothing with nothing. It's impossible to wrest anyone from the grip of dopamine

---

22 C. S. Lewis, *The Screwtape Letters* (London: Geoffrey Bles, The Centenary Press, 1942), 64.
23 Postman, *Amusing Ourselves*, 157.

culture without a more life-giving and worthwhile alternative. We need to pursue (and teach others to pursue) something of substance that satisfies the created needs scrolling videos can only anesthetize. We must think carefully about *why* we find certain distractions addictive and cultivate in their place solid loves for the gifts with which God has filled his world. As we consume—or create—online video in today's world, we might ask questions like these:

- Am I actually enjoying this mindless scrolling, or is it the equivalent of Screwtape's "dead fire in a cold room"? All things being equal, is there something else I would rather be doing right now, and if so, why am I not doing it?
- What God-given gift are time-wasting videos replacing for me? Is it social interaction, friendship, exploration of the natural world, fitness, travel, or something else? How can I pursue the real gift instead of mindlessly consuming videos on my phone?
- What created limits does my body impose on the enjoyment of the gift I desire, and how is online video tempting me to live as if I don't have these limits (think of your need for sleep, exercise, face-to-face companionship, or regular in-person worship)?
- Is there a redeeming purpose to which this medium can be put? What videos have I seen on this platform that genuinely educate, inspire, entertain, or enrich relationships? Have I contributed to or spread that redemptive content in what I've uploaded and shared?
- If I am a Christian creator of online video as a hobby or profession, what can I do—both in the form and content of

my videos—to make sure I'm not just adding to the noise and contributing to dopamine addiction?

- Do I have clearly defined times and "sacred spaces" in which I disengage from screens and pursue other good gifts?[24] Am I giving myself and those in my home sufficient time and opportunities to be human in older ways and "recalibrate" from technology use?

- How can a given use of video encourage creative engagement with the written word, especially God's word? How will I know that it is actually accomplishing this, as opposed to supplanting the written word?

3. Redeem the Medium

This last question is especially important in light of Christian efforts over the years to teach, preach, and evangelize using screens. Postman was far from optimistic about the potential of such projects. He warned that attempts to translate Scripture to television or cinema would result in God's word being subordinated and replaced by the new medium.[25] "There is no doubt," he wrote, "that religion can be made entertaining. The question is, By doing so, do we destroy it as an 'authentic object of culture'?"[26]

Yet consider applications of video like BibleProject, a curriculum of animated YouTube explainer videos hosted by Tim Mackie and Jon Collins, now in its second decade. These videos (none of them TikTok-length) trace themes found throughout Scripture, answer

24 See Sherry Turkle, *Alone Together: Why We Expect More from Technology and Less from Each Other* (New York: Basic Books, 2011).

25 Postman, *Amusing Ourselves*, 96.

26 Postman, *Amusing Ourselves*, 126.

questions arising from the text, and chart the structure of individual books, often in great detail. They incorporate a dialectic style in which Collins anticipates questions and objections viewers might have and Mackie replies.

I've witnessed these videos used to great effect in classroom and small group settings, where they mimic the teaching style of a pastor or seminary professor like the late R. C. Sproul, armed in the 1980s with his chalkboard and boundless enthusiasm. BibleProject may be entertaining and beautiful, but (contra Postman) its main result doesn't seem to be entertainment or dopamine-based distraction. Rather, these videos produce greater biblical literacy and (in my experience) break down barriers of intimidation and confusion that often keep people in our post-typographic age from approaching the biblical text itself.

Or think of another application of video focused on the Bible: the streaming drama series *The Chosen*. This production follows the Gospel accounts through the eyes of Jesus's disciples, filling in fictional details, plotlines, and character development to create what is undeniably (but not solely) entertainment.

While *The Chosen* is conceptually similar to a televised attempt in the 1980s that Postman mocked (*The Genesis Project*), its effort at a plausible portrayal of "the story behind the story" in the New Testament should fascinate anyone curious about video's potential to point us back to reality.[27] After all, the Bible reports only the bare facts of Christ's life. But there actually were backstories, motives, and human factors not recorded in biblical history. As a medium, video seems adept at conveying this, and in so doing, reminds us of an aspect of reality we easily overlook.

---

27 Postman, *Amusing Ourselves*, 96.

4. See Screens as Symbols

Our struggle against distraction requires a willingness to use symbols well and not to mistake them for the realities they symbolize. As I write this, my wife is smiling over at me from across my desk. But it isn't my wife. It's a photograph of her, framed and positioned so I can often glance at it. Few would worry about me confusing this four-by-six-inch symbol with the real woman I married. Yet as we've seen, millions of people now behave as if they've confused videos (which are just rapidly moving images) for the gifts God offers in real life or have accepted videos as cheap substitutes for those gifts. This is a dehumanizing way to live, and it ultimately robs us of the joys for which our senses and our rational, relational souls were created.

Eyes Made for More than Reading

Video, like all other forms of media new or old, long or short, is a symbol—a way of creatively restaging the world. To the extent it distracts us from that world and the God who made us to behold it, we should treat it with caution. But our eyes were made for more than reading, and to deny the viability of perceiving goodness, truth, and beauty *visually* is to deny the goodness of our createdness, by God's design.

Moreover, it should be clear—more than 130 years after the earliest motion pictures—that this captivating way of restaging the world is more than the sum of its frames. Moving mediums, from early silent films to the thirty-second TikTok, all share a power to captivate us and excite bodily responses, precisely because they mimic the real world in a way the written word cannot. Video engages us sensorily, emotionally, and temporally in ways that scripts, screenplays, and even still photographs don't. That's why it so easily distracts. But as often as we watch a video and smile with fondness for the reality it depicts, we prove it's also possible to use this medium well.

Discussion Questions

1. This chapter references a few of Postman's notes on God's scriptural commands against graven images. In what ways have Christ's followers stressed the significance of writing, words, and books from the time of Jesus until now? Is there a biblical precedent for the significance—or even preference—of the written word?

2. Providing several examples of how video has been used well for the sake of the gospel, Morris suggests Christians should creatively redeem video despite its dangers. Are there large groups, churches, or organizations you have seen do this well?

3. At what point (if any) do you think Christians should beware of representing the Bible in visual formats? What would be appropriate or helpful ways to depict the Bible (or biblical concepts) in video formats?

4. Consider the rising popularity of video-based influencers, whether on Instagram Reels or YouTube or other platforms. What are the potential hazards and potential possibilities of Christian "influencers" seeking to leverage this space for gospel mission?

11

Reconnect Information and Action

How to Stay Sane in an Overstimulated Age

Brett McCracken

IN TODAY'S HYPERCONNECTED WORLD, information comes at us fast and furious, from every direction, 24-7. We wake up to news alerts about a major earthquake in Japan or a political assassination in Ecuador. We open our social media feeds and, within the first minute of scrolling, see the latest grim headlines about war or rumors of war, the latest anger-inducing missive in this or that culture war debate, and the latest foolish oversharing from this or that uncle or college friend.

Because we are human and emotionally wired, it's natural that these things provoke us and inflame our hearts to want to *do something*. Yet what can we do with this abundance of troublesome information aside from being informed about it? We are overstimulated but underactivated. Information bombards us but action is elusive. I'm convinced this dynamic is one of the major sources of anxiety and mental unhealth in today's information age, and it's something Neil Postman warned about.

Postman talked about it in terms of what he called the "information-action ratio." For most of human history, there was a high correlation between the information that filled human brains and the tangible actions they could take in response. "News of the world" was inaccessible to most people. The information that concerned them was closer-to-home realities of family, farm, or community: information with direct bearing on the actions of everyday life.

But this all changed, Postman argued, with the invention of the telegraph. Suddenly, the "news of the world" was much more accessible to average people, who found it an amusing novelty. The problem, however, is that this influx of far-flung information "gives us something to talk about but cannot lead to any meaningful action." As Postman observed, "For the first time in human history, people were faced with the problem of information glut, which means that simultaneously they were faced with the problem of a diminished social and political potency."[1]

If Postman's observations about "information glut" were accurate forty years ago, how much more are they today, when we're speeding down the "information superhighway" faster than ever via our ubiquitous smartphones and ever-present Wi-Fi? And the resulting problem of impotence is even more pronounced than it was in Postman's era.

In today's world, it's not just occasional televised traumas that burden our souls; it's the constant feed. "Breaking news" is no longer

---

1 Neil Postman, *Amusing Ourselves to Death*, 20th anniversary ed. (1985; repr., New York: Penguin Books, 2005), 68.

the alarming verbiage that signals a rare calamity; it's the everyday parlance of twenty-four-hour news and social media publishers skilled at the art of clickbait. These media publishers are eager to garner eyeballs by any means necessary. Another school shooting. A salacious scandal. An election "shock poll." A helicopter-filmed police chase. An Amber Alert for a missing child.

But what are we to do with all these alarming headlines and triggering dings of "breaking news"? Media outlets don't care about this question. Their only interest is that we have tuned in, clicked, and fallen for the pseudo urgency of the Important Information they've put on our radar. Making audiences "aware"—at best, helping them become "informed citizens"—seems to be the chief value proposition the news industry can offer in its defense. But awareness to what end? Is this tidal wave of chaotic information informing us merely for the sake of us "being informed"?

Awareness as an End unto Itself

We've come to a point where, yes, the primary goal of most information mediated to us is that we should be informed and aware of it. Not *educated* or *activated* about important things happening in the world, mind you; merely *aware*. The benefits of an informed citizenry have long been trumpeted as a valorous purpose of the free press (and indeed, the benefits are real). But we also need to talk about the liabilities that come with an *over*informed or *trivially* informed citizenry.

In *Amusing Ourselves to Death*, Postman argued that TV had altered the meaning of "being informed" by "creating a species of information that might properly be called disinformation." This is not the same as outright misinformation, he said. It's rather *misleading* information, which "creates the illusion of knowing something but which in fact leads one away from knowing."[2]

In Postman's view, mass media (led by television) created a world of dilettante experts whose absorption of vast amounts of information—packaged to them as entertainment—gave them a false sense of know-how about the happenings of the world. Referencing this know-how (e.g., "I saw this news story about ____" or "I read this *Atlantic* article about ____") became status markers. Information *awareness* took on a cultural cachet quite apart from its actionability.

Fast forward four decades, and we now take it for granted that "awareness" is a value in its own right. The conversation starters might be different today ("I saw this TED Talk on YouTube about ____" or "I saw this TikTok about ____"), but the status it brings has only increased. Our ability to cite, allude to, or summarize secondhand information about a breadth of things (even if our grasp of the "thing" is actually wafer thin) turns information into a means of signaling our claim on that most coveted virtue, *relevance*.

For digital natives who've lived their whole lives in a hyperaware, globally connected information ecosystem, it's understandable that a word like *woke* would come into prominence as a shorthand for social justice. In the twentieth century, social justice "activism" involved tangible actions like volunteering or picketing in a real physical place; in the twenty-first century, someone can be an "activist" without ever getting off his or her phone. Activism (or "slacktivism") moves from being primarily about *doing* to largely about *saying*: participating in the correct lingo, hashtags, and accepted speech (e.g., preferred pronouns) becomes the means of activism more than, well, *actions* offline. "Doing justice" becomes a discursive activity more than a tangible one.

---

2 Postman, *Amusing Ourselves*, 107.

In this upside-down world, people can—and often are—accused of apathy and inaction for being silent ("silence is violence") on social media, even if their offline, unpublished activities are thoroughly oriented around addressing the injustice they're being accused of ignoring. So it goes in a world where discourse about a problem (talking about it publicly) occupies a higher social standing than actual efforts to solve the problem.

This is problematic.

Problem of Being Overinformed

Five years after publishing *Amusing Ourselves*, Postman gave a speech to the German Informatics Society that elaborated on the information-action ratio. In the talk, titled "Informing Ourselves to Death," Postman described how, for the average person in 1990, "information no longer has any relation to the solution of problems." The way he described it could just as easily describe the average person in 2025:

> The tie between information and action has been severed. Information is now a commodity that can be bought and sold, or used as a form of entertainment, or worn like a garment to enhance one's status. It comes indiscriminately, directed at no one in particular, disconnected from usefulness; we are glutted with information, drowning in information, have no control over it, don't know what to do with it. . . . Our defenses against information glut have broken down; our information immune system is inoperable. We don't know how to filter it out; we don't know how to reduce it; we don't know how to use it.[3]

---

3   Neil Postman, "Informing Ourselves to Death" (address to the German Informatics Society, Stuttgart, Germany, October 11, 1990), https://web.williams.edu/HistSci/curriculum/101/informing.html.

Remember, Postman observed this "information glut" problem in the pre-internet era. How much more are we glutted with information today? If we didn't have good "information immunity" defenses back then, we're even worse off now—especially in the age of ChatGPT, deepfakes, political misinformation campaigns, and the resulting epistemological crisis. The information crisis we face is at least threefold: *too much* information that moves *too fast* and is algorithmically tailored to be *too focused on me.*[4] In a sense, "being informed" is more of a liability than an asset in today's world. The quality of digitally mediated information is simply too untrustworthy.

What happens to us when we're overinformed but underactivated? From my experience and observations, some common side effects occur.

We become anxious. When a world's worth of "breaking news" calamities, injustices, and apocalyptic headlines steadily feed our souls, we naturally feel anxious and on edge.

We become angry. Rising blood pressure and seething anger follow when we're constantly exposed to partisan clickbait, triggering troll provocations, and other forms of foolish talk.

We become addicted. Algorithms easily figure out what types of information each of us can't resist. Soon we're scrolling and clicking like addicts, unable to resist the intoxicating allure of our favorite genres of "news," trivia, or juicy gossip.

We become numb. A diet of information disconnected from tangible action makes information abstract and surreal, disconnected from our real life. Eventually, headlines about a horrific mass shooting become things we scroll past as casually as we glance at a friend's vacation photo.

---

4 I devote chapters to each of these three challenges in my book, *The Wisdom Pyramid: Feeding Your Soul in a Post-Truth World* (Wheaton, IL: Crossway, 2021), chaps. 1–3.

We become lonely. When we spend large segments of our lives binging on digital information far removed from local, embodied communities—even if it's information we debate or discuss with others online—we become lonelier. The online influencer we listen to, or the interlocutor avatars we fiercely debate, are hardly substitutes for the know-and-be-known community we really need.

We become delusional. Because of the algorithmic shape of information today, no two of us live in the same information universe. We all see things differently, in ways tweaked to please our preferences and biases. Naturally, this further entrenches us in echo chambers, deepening our confidence in our own rightness (however wrong we are).

We become detached from reality. The cumulative effect of all the above is that an overinformed life becomes a pseudo real life. When awareness trumps action and we're more compelled by narratives than by reality, our sense of the world becomes ever more surreal.

Perhaps C. S. Lewis sums it up best in this letter to a friend, when he laments the dynamics of an information-action disconnect:

> It is one of the evils of rapid diffusion of news that the sorrows of all the world come to us every morning. I think each village was meant to feel pity for its own sick and poor whom it can help and I doubt if it is the duty of any private person to fix his mind on ills which he cannot help. (This may even become an escape from the works of charity we really can do to those we know.) A great many people do now seem to think that the mere state of being worried is in itself meritorious. I don't think it is.[5]

5   C. S. Lewis, letter to Dom Bede Griffiths (1946), quoted in Paul F. Ford, ed., *Yours, Jack: Spiritual Direction from C. S. Lewis* (New York: HarperCollins, 2008), 119.

Not only is Lewis right to challenge the social merit attached to "the mere state of being worried" (i.e., the social capital of *awareness*), but he hits the nail on the head when he says we should avoid fixing our minds on problems we can't solve. This not only burdens us in all the ways described above but tends to distract us from the local problems we *can* help fix.

Neglecting the Local

With all the energy we devote to keeping up with the goings-on of the world, we might neglect the people we can love and the problems we can address in our own backyards. For Christians called to love our neighbors and tangibly pursue mercy and justice, this is the crux of what's wrong with an imbalanced information-action ratio.

Such is the state of our mass-mediated information environment that your average twenty-first-century young person can tell you far more about national politics than local politics. He develops strong opinions about presidential candidates and Supreme Court cases but couldn't tell you the name of the mayor or a city council member in his city, nor identify the most pressing challenges facing his proximate community.

Of the millions of Gen Zers who posted a blank black square on Instagram in June 2020 (#blackouttuesday) to protest police brutality, how many have ever had a conversation with a police officer in their own neighborhood? Of the millions who changed their social media avatars to the Ukrainian flag in February 2022, how many have tangibly helped refugees or immigrants from war-torn nations in their own cities?

Online hashtag actions are well intentioned. And maybe the viral power of such "collective online action" makes some difference. But as Lewis points out, the danger is that such actions "become an escape from the works of charity we really can do to those we know."

There are many reasons why everyone should strive for a more balanced information-action ratio. It'll help your mental health and ground you in local life and embodied community. For Christians specifically, it'll remind you of your creaturely limits and deepen your trust in a sovereign God who is omniaware in ways you can never be. And it'll present more fruitful avenues for loving your neighbor and being a faithful witness in the particular place where God has situated you.

Bringing Balance to the Ratio

Christians should be countercultural by striving to reconnect information and action, modeling a healthier way of living for a world out of balance. How can we do this? Here are ideas for individual Christians and ideas for churches and leaders.

For Individual Christians

Audit your news and information diet. Make intentional efforts to reduce your intake of national and global information while increasing your intake of local information (which has more potential to be actionable). Don't turn your ears off to the cries of the world. But listen more eagerly to the cries closer to home.

Embrace your limits. As you become more "unaware" of the steady hum of information in the news that might be making others anxious, angry, and stressed, see this as an opportunity for resting in God's sovereignty and praising him for his power. A world's worth of burdens is too much for you—but not for God. Contemplating our limits in contrast to God's unlimitedness is a fruitful path toward wisdom (see Ps. 90).

Rejoice in how God designed you. You are an integrated mind and body. What comes into your brain has a natural outlet in your physical activities. You weren't made to just be aware of faraway problems and global chaos about which you can't do much. You were made to bring order to the chaos in your immediate vicinity. You weren't made to be a gawker but a gardener (Gen. 2:15).

Pray. Prayer is an important action we can take. When you inevitably encounter information about an injustice or tragedy in some far-flung part of the nation or world, don't let the information sit idly in your troubled brain. Take it to the Lord in prayer. As much as our secular culture demands more than "thoughts and prayers," Christians know prayer is actually potent and crucial. If we can't do anything else in response to troublesome information, we can pray to the one who can.

For Churches and Church Leaders

Disciple people in media habits. Information intake should be a subject addressed in discipleship—not in a legalistic sense but as part of wisdom. Help the people in your church think through the amount and type of information they consume and how it's shaping their souls.

Promote localism. Church leaders should lead people (especially Gen Z and Gen Alpha) to prioritize the local, proximate, and offline as much or more than the distant, disembodied goings-on of the online world. Make the case for why a balanced ratio of information and action is not only a recipe for improved mental and spiritual health but conducive to a more effective Christian mission.

Gather people for prayer. When some national or global calamity does occur, in such a way that most in your Christian community will be aware of and troubled by it, prayer is an appropriate communal response. Both in the regular church gathering and in impromptu meetings, the church can and should take the action of prayer. It's an "action" in the truest sense, and one we should never neglect.

Call people to take action. Churches should regularly organize opportunities for people to tangibly solve real problems in the community. Often this works best by establishing long-term partnerships with organizations already doing specific work that aligns with biblical neighbor love: crisis pregnancy centers, foster and adoption agencies, homeless shelters, food distribution centers, and so forth. There is no end to the needs in your own backyard. And if a national news headline happens to be about something happening in your city or community, then your church should spring into hands-on action. This is a rare opportunity for burdensome information about calamity to directly translate to tangible community service, in partnership with local organizations and civic authorities.

Beauty of Activated Church

For much of my adult life, I was an overinformed news junkie. The onset of social media amplified this addiction—and my soul suffered as a result. Thankfully, I found a healthier way to live, in no small part because I rediscovered the beauty and necessity of the local church.

Once I gave myself wholeheartedly to local church life, I came to see that the burdens and griefs of ten people in my small group were far more important for me to carry than the burdens and griefs of countless sufferers on social media. Not only could I see the actual tears on actual faces as they shared, but I could hug them and know them in their suffering—and help them *through* it.

I also came to see that the *tangibly activated* local church is a far more satisfying and functional community than the *virtually aware* community of social media. Whether they're distributing food in partnership with local food banks, mobilizing volunteers for a local foster and adoption agency, or simply rallying the congregation around the needs of the community (single moms, meal trains for sick families, house cleanup for elderly members, and so forth), a church's localized, tangible action is beautiful to behold.

And when troublesome news from distant places does reach our corner of the world—as it invariably will—the local church is where I go first to process and pray through it, even if no other "action" is possible in response. For centuries, the church's "prayers of the people" liturgies have borne witness to the fact that in those instances where we can't "do" anything with our hands to help, we can always drop to our knees and pray.

Christians can model a different mode of living in an over-informed, underactivated world. It's a mode that isn't numb or ambivalent to the countless problems that plague our world but realistic about our limited scope and where we can best be used. It's a mode that leads to calmer minds, more focused souls, and more engaged bodies. It's a mode that syncs up with how we were created and resists the digital era's many temptations toward god-like limitlessness.

Discussion Questions

1. What aspects of the digital age have led us to a place where *talking* about a problem is more common (and arguably more compelling) than *solving* a problem? What assumptions inform the slogan "silence is violence," and how should believers respond to this accusatory statement?

2. Take a look at McCracken's list of the side effects from being "over-informed but underactivated." Which of these do you find most prevalent in your experience?

3. How do certain abilities of a smartphone condition us to believe we have "godlike" qualities such as omniscience, omnipotence, and omnipresence? How might we embrace our creaturely limits as a way to push back?

4. McCracken highlights "relevance" as a "coveted virtue" in our digital society. How has this core cultural value affected the ways we talk about Jesus in the church and in our missional efforts? What are the dangers of an inordinate focus on relevance?

12

Embrace Your Mission

Tangible Participation, Not Digital Spectating

Read Mercer Schuchardt

THE PURPOSE OF A COLLEGE EDUCATION, Neil Postman used to tell us in graduate school, was to win television game-show contests; it is the only context in which all that disconnected trivia you've picked up in four years could have a possible use. The fact that this "use" was a ruse, and that this context was a "pseudo context," was something we had to figure out on our own if we weren't already in on the joke.

Having never owned a television in either childhood or adulthood, I thought it was pretty funny because it seemed to be the most plausible explanation for the "use" to which I could put all the random information I'd garnered majoring in comparative religion, I mean French, I mean English literature by the time I'd stretched four years into six at Swarthmore College.

But today? Right now? What is the purpose of fragmented, irrelevant, context-free information in the internet age? It is what T. S. Eliot said in the 1930s: to distract us from distraction with distraction. As I write, one current use is to sell Trader Joe's mini tote bags for $230 on eBay after buying them at $2.99 from the local grocery store, but only after posting a seven-second TikTok video to show how adorable they are. In other words, the new purpose is to play the digital media algorithm for and against your audience simultaneously to become a profitable social media influencer. One victim actually said this out loud: "Did I need this? No. Am I gonna use this often? Probably not, but it's so [expletive] cute I had to have it."[1]

What would Jesus do? This is an increasingly useful question to ask in the digital age, even if you're not a Christian.

Would Jesus have a Facebook page? Would he keep a running Snapchat streak with his disciples? Would he post his political opinions on X? Would he take a picture of himself from above after a particularly grueling day of lifting rocks for his stonemasonry job and post it to Instagram, knowing that "jacked" preachers are more appealing to the young? Would he take out Super Bowl ads telling us that "he gets us"? Would he scroll through TikTok while listening to Spotify in order to know the mindset of his audience, who themselves had become unwittingly but empirically micro-niched by demographics, psychographics, and other quantifiable "vibes" of the marketplace of goods, ideas, attention? Most importantly, which color of mini tote from Trader Joe's would he rock?

---

1 By the time you read this, the TikTok influencer hype of Trader Joe's mini totes will have been long eclipsed by a thousand or a million subsequent pseudo contexts, but at the time of writing you could read about it in *Business Insider*. Grace Dean, "Shoppers Are Scrambling to Get Their Hands on Trader Joe's $3 Mini Tote Bags, Leading to a Lucrative Online Resale Market," *Business Insider*, March 11, 2024, https://www.businessinsider.com/.

This is just another version of asking a bigger question: How would Jesus mediate his relationship to his followers if he were to come back today? My view is that he would communicate in the year 2025 the exact same way he did circa 25: live, embodied, in-person, using just his voice to speak within the range of unamplified earshot and his hands to heal by touching those who needed it. He would speak most often and most commonly to individuals, occasionally to large groups no greater than five thousand, and in general would let people find him as much or more often than he went to find them.

I may be wrong. But even if I am mistaken in this conjecture, I suspect if you were to imitate Jesus as an antidote to the digital distractions of 2025, you'd discover you were a lot less "busy," a lot more "relevant," and about 80 percent more "influential."

All this is to say that what the world needs from Christ and his followers is not another logo, another website, another influencer trend, or another social media app to compete in a world of 8.93 million apps.

What the World Needs from Christians

What the world needs from Christians is for them to embrace the tangible missions God commanded them to do and that are all still needed: (1) tend the garden (Gen. 2:15); (2) be fruitful and multiply (1:28); and (3) go tell all nations the gospel (Matt. 28:19). Find a job and work hard to bring order out of some genre of chaos. Find a spouse and build a family. Find ways to love your neighbors and spread the gospel. Feed the poor. Care for the orphans. Tend to the sick. You know— those unglamorous things Christians have built their reputation on throughout history.

Whatever your vocation, and however you've been gifted to live out God's call on your life, you don't really need the distractions, viral causes, and virtue signaling of social media. They add little substantive value to the lifelong labor of love you give to, say, 150 people in your life.[2] Heed the oldest vocational advice in the New Testament: make it your ambition to lead a quiet life, work with your hands, and be dependent on no one (1 Thess. 4:11–12).

What does a meaningful, missional, light-in-the-darkness, embodied presence look like for Christians in a digital world? Perhaps it looks like visiting your grandmother in the retirement home rather than Zoom-calling her or visiting the open floor of a hospital ward and praying for those who accept your offer of prayer. Maybe it looks like getting a group of investors together, buying a defunct church, and not turning it into an upscale private condominium complex but rather renovating it to be a new, lively church that preaches gospel life and seeks to meet spiritual and material needs in your twenty-mile radius: a food pantry, a community garden, a diaper bank, or a volunteer-based daycare for one-parent families or families in which both parents have to work. The possibilities are nearly endless.

What such a life doesn't look like is getting sucked into the all-consuming screen world that—if we're not careful—gradually draws us out of the real and into the surreal, out of embodied rhythms and relationships and into endless scrolling and avatar abstractions.

---

2 The Dunbar number refers to how many stable social relationships a human can maintain. According to British anthropologist Robin Dunbar, 150 is the maximum amount. See "Dunbar's Number: Why We Can Only Maintain 150 Relationships," BBC, October 9, 2019, https://www.bbc.com/.

Life Abundantly

Between TV and digital media, the average American uses screens for 10.85 hours per day.[3] Name any other activity, in any other age, that people of any other tribe, nation, ethnicity, religion, gender, or sexuality ever did for 10.85 hours per day. You can't work for that long, you can't sleep for that long, you can't eat for that long, and you can't make soup for that long. Maybe the only thing you can do for that long is play the online fantasy game Prius for 10.85 hours straight, in which you raise a virtual baby you've named Anima, while your own four-month-old human baby dies from neglect.[4] So while it sounds hyperbolic and extreme to claim the digital world is literally "killing you"—these are, thankfully, very rare cases—it is the case that digital life is robbing you of your lived, embodied, real, and valuable life.

It's no surprise that devoting 10.85 hours of your sixteen-hour waking life to nonreality has negative consequences in the real world. It's no surprise that rates of an array of adverse health effects, including body/reality/temporality dysmorphia, go through the roof. We have never lived in an age more anxious, lonely, depressed, and suicidal than the one we are in right now—because history has never seen one. This is the zombie apocalypse, this is insanity, this is death, and this is death more redundantly.

And what, by comparison, was Jesus's promise? That you would have *life*, and life *abundantly* (John 10:10).

At a certain point, everyone asks some version of these questions: What is worth doing? What remains? What can I do in this life

---

3 A. Guttmann, "Media Usage in the U.S.—Statistics & Facts," Statista, December 18, 2023, https://www.statista.com/.

4 Andrew Salmon, "Jail for Couple Whose Baby Died While They Raised Online Child," CNN, May 28, 2010, https://edition.cnn.com/.

that will matter to those who come after I am gone? These are the outgrowths of foundational questions all religions try to answer: Who am I? Why am I here? And what should I do? One of my students put it eloquently in a recent essay: "Who am I? Why am I here? And why don't I care?" The title of philosopher Mark Kingwell's book on boredom is telling: *Wish I Were Here.*

Here's an answer: you should avoid the pseudo context the smartphone produces and avoid "hashtag causes" like the plague. If you question the value of such causes, here's a simple test: Name the Ukrainians you personally helped by posting their national flag on your social media page. Name the Russians. Name the Palestinians you helped by saying you were against the war. Name the Israelis.

Real love requires real sacrifice. It's proof of social media's weaponization of your empathy that you "think" you did a good thing because it "feels" like you did a good thing. But if you can't name the individuals you helped, or say how you helped them, you likely did nothing except make yourself feel good. This is the demonic brilliance of most social media platforms: they make you feel like you're participating in a justice cause, developing a meaningful community, or making a suffering person's life easier (by "liking" her Instagram post). But mostly what you're feeling are vibes more than realities. And we were created for realities, not vibes.

The chief and highest end of man is to love God and fully enjoy him forever.[5] And part of enjoying God is giving ourselves to full participation in the *process* of life, because life itself is a good gift from a good God. But digital media's distractions are quietly converting us from participants in the process of life to spectators of other people's lives.

5 Westminster Shorter Catechism, q. 1.

Here's another answer: you could imitate the Amish because they are quietly and meekly not losing their way.

Right now, the average American family consists of 3.13 people, the average Muslim family 6.4 people, and the average Amish family 10 people. The Amish went from a population of 125,000 to 250,000 in just twenty years, with a new Amish settlement founded every three weeks.[6]

The more people at your party, the better the party. While some people are spending their twenties scrolling up and down, swiping right and left, and generally looking unidirectionally at their devices, others are busy getting married, having babies, and "tending the garden" in a way that gives more than it takes. Did you know that the people who have the most people tend to win the heart and soul of any culture? The future has always belonged to those who show up for it. Demographics is not just economic and political destiny; it tends to be a fairly accurate predictor of spiritual destiny as well.

You can get in trouble for saying this out loud to a class full of undergraduates. But one shockingly tangible mission you could embrace that would contribute to the kingdom would be to marry young, have a larger-than-average family, and raise them in the nurture, admonition, and fear of the Lord. That may do more than any other missionary activity ever will or could, simply because of the fact that one out of five Christians is a convert from another religion[7] (or no religion) while the other four out of five are a result of being born into a Christian family and being raised from their mothers' milk of belief to the solid food of true discipleship. The larger your base,

6   Timothy Aeppel, "The Amish Population Boom," *Wall Street Journal,* July 29, 2010, https://www.wsj.com/.
7   "Modeling the Future of Religion in America," Pew Research Center, September 13, 2022, https://www.pewresearch.org/.

the more you can afford a "black sheep" who leaves the fold, which is one reason the Amish of Holmes County, Ohio, don't sweat their losses in a religious community with a retention rate of 95 percent.[8]

This Is Not a Vibe

There is no app for this. This is not a vibe. There is no shortcut to deferred gratification, which is the very essence of faith. The etymology of the word *religion* (according to Cicero) is *relegere*: to reread. Purpose isn't found in our awareness of TikTok lingo and the pseudo events of digital discourse. It's found insofar as we serve the God of history—the God who is outside of time and who gave us speech in the first place.

The advice I offer here is the same advice I preach to myself: work out my salvation with fear and trembling (Phil. 2:12). Visit grandma. Bring shared meals to my elderly neighbors. Walk across the street to meet my neighbors and invite them over for a meal. Call my wife more often from work and tell her how thankful and grateful I am for our life together. Wash the dishes. Go for more walks and bike rides. Remember my grandchildren's birthdays and send a card if I can't afford a gift. Pay closer attention to my inattentive students. Fast. Pray. Tithe.

See if you can be a participant in life more than a virtual spectator of other people's lives. Try to understand how much media influences your perception of reality and how often it warps, distorts, or betrays the truth. Take Eric Brende's advice and become a "minimite" by divesting yourself from the corporate tentacles that have

---

8   Amish retention rates vary by community, with 80 to 90 percent being the norm, and 95 percent being the highest recorded so far, among the Andy Weaver Amish of Holmes County, Ohio. Erik Wesner, "When Do Amish Get Baptized? (All About Amish Baptism)," Amish America, https://amishamerica.com/when-do-amish-get-baptized/, accessed June 28, 2024.

entrapped nearly the entire spectrum of your political, economic, and spiritual life.[9] Read between the lines of the spirit of the age and see if you can muster the courage to quietly fight back. Resist any cultural force that would seek to turn your family tree into a stump. Obey Jacques Ellul's dictum that existence is resistance, which admittedly sounds more elegant in French: *L'existence, c'est resistance!* Find the line between freedom and necessity, and walk that line.

You were not created to be a predictably programmed series of increasingly addicted responses to the stimuli of digital dopamine delivery systems. You were definitely not called to create the Christian version of those demons. You *were* called to be sober, alert, and watchful, to study the signs of the times like the men from Issachar (1 Chron. 12:32) and thereby to be the active agent of change in the world, to be the stimuli the world responds to—which means to be the salt, yeast, and light that brings the flavor, raises the dough, and dispels the darkness.

As Ellul puts it, Christians are supposed to be "troublemakers, creators of uncertainty, agents of a dimension that is incompatible with society."[10] When did we lose our verve?

Postman's One and Only Email

Neil Postman taught me the most, without intending to, by both the form and the content of the one email he ever sent. He sent it in 1997, the year the internet became a mass medium. Its content is so astonishing, so funny, and so insightful that you will find it almost prophetic, because this was written before any social media existed and a decade before the iPhone was invented.

---

9   See Eric Brende, *Better Off: Flipping the Switch on Technology* (New York: HarperCollins, 2004).
10 Peter Steinfels, "Jacques Ellul, French Critic of Technology, Is Dead at 82," *New York Times*, May 21, 1994, https://www.nytimes.com/.

He wrote it in response to the creation of an academic listserv through which his graduate students could hold an online discussion of their studies, a purpose most people today would agree was a "good use" of the new medium. Here is the one and only email Postman sent in his seventy-two-year lifespan:

Subject: Observing the Law, 1997

This is the Ghost of Marshall McLuhan speaking to you. I don't have to tell you from what world I come. I am using Chris Nystrom's facility in order to reach you. I will say what I have to say only once. You will not hear from me again unless you persist in your foolishness.

Does the word "books" mean anything to you? Do you have so much time on your hands that you can afford to waste yourselves on this infernal machine? Have you already accumulated so much wisdom that you no longer need to read the best that has been thought and written? Is this the way you honor the work and life of my great friend and disciple, Neil Postman? Do any of you actually know how to spell?

I have now read all of your idiotic messages. Hear, now, The Law: Every medium taken to its furthest extent flips to its opposite. Thus the written word, which is the source of all the intellect we have, when used in this unholy fashion becomes a medium for the expression of all our stupidities. This, you have demonstrated amply. Enough, I say.

I must now return from whence I came. Remember what happened to the Hebrews when they did not follow the Law.

—*Ghost*

"A medium for the expression of all our stupidities." Is there a better description of the actual eternal value of social media than that? So here I am, Postman's disciple, letting you hear from Postman again since, clearly, we have persisted in our foolishness to an even greater degree than he could have anticipated in 1997.

Go touch grass. Put down the phone, give up the screen, and initiate: no matter your age, stop scrolling and start your life. You need only ten thousand hours of deliberate practice to get good at something worth doing, and you've got that in spades if you give up the 10.85 hours per day currently devoted to media. That's just 749 days to get really good at your skill, art, trade, or craft; that's just two years, which is half the time it takes to acquire a college education.

Stop trying to reach the 150,000 impressions per video required to be an influencer in the virtual world. There are 150 people in your real world whose lives you could tangibly influence for eternity. Life is not a popularity contest; that's your worst high school nightmare. Life is a constant invitation to accept the challenge of being the presence of the kingdom, here and now, for those around you. Christ changed one life at a time. Christianity grew by believers following his model. Go and do likewise.

One day in graduate school, Postman invited me to have lunch with him in the faculty lounge at NYU, and we had one of the most delightful, enjoyable, and entirely forgettable conversations I've ever had. The only thing I remember was that when we got back to the Department of Culture and Communication, he paused at the door with that name on it and said to me, "We called it the Department of Culture and Communication because we wanted it to include everything; it would be like calling it the Department of Man and Woman." Indeed, "culture and communication" encompasses pretty much everything that shapes us—but so much of it is invisible until someone points it out. My colleague Arthur Hunt ended his tribute to Postman with this advice for Christians: "Like Postman, we should not be afraid to be counted among those who raise their voices, as he said, 'to a near-hysterical pitch, inviting the charge that they are everything from wimps to public nuisances to Jeremiahs. But they do so because what they want others to see appears benign, when it is not invisible altogether.' "[11]

Digital media has played its hand, and by almost all quantifiable measures its primary effect has been to perpetuate despair in the human soul.

What was the most successful culture-and-world-changing strategy in human history? It was the embodied speaking and healing-by-laying-on-of-hands that Christ, his disciples, and apostles practiced. That won't change. Do you want to imitate Christ? Do you have a voice that can speak? Do you have hands that can help? Do you have the will to put the phone down?

Get going.

The people who need the Christ within you are within earshot. You can probably see them from where you're sitting.

Discussion Questions

1. Robin Dunbar finds that 150 is the maximum number of stable social relationships a human can maintain. How does contemporary technology break that barrier? Where in your life have you seen the dynamics of online life giving you too many relational connections?

11 Arthur W. Hunt III, "Remembering Neil Postman: A Media Legacy Christians Should Consider," *Reformed Presbyterian Witness*, November 6, 2005, https://rpwitness.org/.

2. "[Jesus] would communicate in the year 2025 the exact same way he did circa 25." Do you agree? And if so, what does that mean for how we—as Christ's followers and as evangelists—communicate the gospel today?

3. How has our screen-addicted, disembodied, fragmented existence caused us to miss "life abundantly"? Where do you feel this most in your experience of life in the digital age?

4. What in this chapter might give Christians permission to live less hurried, less anxious, and less despairing lives? Consider specific ways you can implement Schuchardt's advice in your life.

13

Cling to Embodiment
in a Virtual World

Jay Y. Kim

NEIL POSTMAN DROVE A HONDA ACCORD. He was fond of telling an anecdotal story about the day he purchased the car. The salesman pitched him on the necessity of cruise control. Postman asked, "What is the problem to which cruise control is the answer?" The salesman replied, "It is the problem of keeping your foot on the gas." Postman responded that in several decades of driving, keeping his foot on the gas had never been a problem. He purchased the Honda Accord anyway because, "as it turns out, you cannot get a Honda Accord without cruise control."[1]

In a lecture given several years after the publication of *Amusing Ourselves to Death*, Postman proposed a list of questions we must

---

1   Neil Postman quoted in "What Is the Problem to Which Electric Windows Are the Answer?," Afflictor.com, April 10, 2012, https://afflictor.com/.

ask of any technology.[2] What he asked the car salesman and a follow-up question both stand out as particularly important for church leaders seeking a more faithful ecclesiology in the digital age.

What is the problem that this new technology solves?

What new problems do we create by solving this problem?

Cruise control on a car is fairly benign. Postman's anecdotal story is, on the one hand, funny because cruise control on a car isn't of any serious consequence. But when the same questions are applied to ecclesiological conversations about "digital ministry," "online church," or other disembodied substitutes for the gathered people of God, the implications become significantly weightier.

What is the problem that digital ministry and online church solve?

What new problems do we create by solving this problem?

Church in the Digital Age

In the spring of 2020, as a global pandemic thrust us into separation and isolation, churches responded by going online. For the next couple years, churches were confronted with a question that had been latent beneath the surface of our pragmatic ecclesiology for more than a decade prior: What does it mean to be the church in the digital age? During the pandemic, the problem that online church solved was clear. At a time when gathering in person wasn't possible, digital platforms offered a chance to exchange pixelated presence and receive content via screens.

This problem no longer exists in most parts of the country and world. Gathering is possible. But new problems were created by our solution to the previous problem. Problems that had been simmering beforehand were brought to a raging boil by our

---

2 Neil Postman, "The Surrender of Culture to Technology" (lecture, College of DuPage, March 11, 1997), YouTube video, June 3, 2013, https://www.youtube.com/.

digital solution—namely, the problems of *convenience* and *low commitment*.

During the pandemic, "gathering" online was a concession. But concession turned to convenience. Though we were able to gather again, many found it easier—on their schedule, comfort, and need for leisure—to stay online.

During the pandemic, even the most ardent introverts were grieved by the absence of an incarnate community. So we committed to logging on. But that commitment tapered. In many of our churches, showing up online when it was the only option was *high commitment*. Showing up online, however, when it is an easy option is an expression of *low commitment*.

These are the new problems the church faces in the digital age. And they're undoubtedly problems for two reasons:

1. Meaningful connection is always inconvenient.
2. Meaningful community always demands high commitment.

Expanding on his once-teacher Marshall McLuhan's concept of "rear-view mirror thinking," Postman explained that to assume a new medium is merely an extension of a previous iteration is foolish and dangerous. With television, for example, "to make such a mistake . . . is to misconstrue entirely how television redefines the meaning of public discourse. [It] does not extend or amplify literate culture. It attacks it."[3] It's no exaggeration to suggest that a continued emphasis on "taking new digital ground" may not actually be extending or amplifying the church's mission but attacking it.

3 Neil Postman, *Amusing Ourselves to Death*, 20th anniversary ed. (1985; repr., New York: Penguin Books, 2005), 83–84.

Some nuance is necessary. In the years since lockdowns lifted, I've met new people at our local church on an almost weekly basis. All those I've asked have told me they watched a worship gathering online before attending in person. Proponents of online church may be correct in their claim that the internet offers churches a "front door" unlike any other in history. But when we build homes, though we may give some energy to the style and color of the front door, exponentially more time and money is spent on the more vital parts of the dwelling.

A beautiful front door or an entirely stunning exterior can make a home nice to look at, a monument to observe and admire. But it's the quality of the interior that determines the home's ability to host well and turn physical space into a relational place of belonging. The former emphasizes a viewing experience while the latter cultivates the opportunity for much more intimate connection. We exchange rushed hellos and sign for packages at the exterior of homes; we break bread, build friendships, share life, and lend our presence within the interior of homes.

Online church is a viewing experience. But the Bible's primary metaphor for the church is family. Countless times, followers of Jesus are called "brothers and sisters," which in the first-century world was no small thing. The familial bond between siblings who shared the same father was one of the strongest social bonds in the ancient world. For the first Christians, conceiving the church as family was a radical reorientation of priorities. It was an utterly inconvenient and almost impossibly high-commitment mandate.

Sibling relationships are messy because they're honest and bare, even brutally so. There's no hiding, mask wearing, or pretense. They're particularly challenging because siblings aren't chosen. But it is this very inconvenient, high-commitment calling that most effectively shapes who we are and who we become as God's people. This transformation of unlikely, disparate, otherwise incongruous individuals into a family does not and cannot take shape at front doors alone, where we exchange only pleasantries and polite, half-hearted smiles.

Reflecting once again on television's pervasive reshaping of American culture in the late twentieth century, Postman wrote that entertainment had become "the natural format for the representation of all experience. Our television set keeps us in constant communion with the world, but it does so with a face whose smiling countenance is unalterable."[4] The "constant communion" that television once offered now pales in comparison to the pervasive, ubiquitous connectivity of our smartphones, tablets, and devices. And with each click, swipe, and scroll, we encounter yet another series of faces "whose smiling countenance is unalterable," offering us polite pleasantries and robbing us of the messy, painful, beautiful, and transformative gift of meaningful community as the embodied church.

Opening Doors

"I believe I am not mistaken in saying that Christianity is a demanding and serious religion," Postman mused. "When it is delivered as easy and amusing, it is another kind of religion altogether."[5] Written in the era of televangelists, Postman's insight demands our attention anew today, in an age of what one *New York Times* journalist calls "Instavangelists." These influencers offer their followers a chance to "worship at any time of day or night at the electric church of [their] Instagram feed" without having to put down their

4 Postman, *Amusing Ourselves*, 87.
5 Postman, *Amusing Ourselves*, 121.

phones.[6] This is easy, amusing Christianity. But there is nothing easy or amusing about Jesus's invitation: "If anyone would come after me, let him deny himself and take up his cross and follow me. For whoever would save his life will lose it, but whoever loses his life for my sake will find it" (Matt. 16:24–25).

Most church leaders aren't looking to become "Instavangelists." Still, the on-demand nature of digital platforms tends to work against a cross-carrying life of self-denial. Christ's offer of transcendent joy comes by way of a path diametrically opposed to the ease and amusement of consumer culture. Jesus, "for the joy that was set before him endured the cross" (Heb. 12:2). In God's kingdom, joy isn't found on a recliner, watching online church in between a Netflix series and a YouTube video. Joy in God's kingdom comes by way of high commitment to the inconvenient mess of embodied communities awkwardly learning to share life together in all its pain and joy.

The apostle Paul writes in Ephesians 2:19, "You are no longer strangers or aliens, but you are . . . members of the household of God." The Greek word for "strangers" (*paroikoi*) and the single Greek word for the phrase "members of the household" (*oikeioi*) share the same root word—*oikos*, meaning "home." A "stranger" is one who dwells *near* the home, while "members of the household" are those who dwell *in* the home.

In recent decades, many churches moved away from "membership," fearing it suggested exclusivity as a sort of Christian country club. But a few years ago, our church returned to it in both language and practice. We wanted to invite people to belong as family members—not casual attendees who consume when convenient

6 Leigh Stein, "The Empty Religions of Instagram," *New York Times*, March 5, 2021, https://www.nytimes.com/.

but brothers and sisters who contribute because we belong. To our joy, we've discovered that the invitation to meaningful commitment has been received as a breath of fresh air by most, a reprieve from the come-and-go-as-you-please, noncommittal nature of online communities. The digital age has disconnected us via ease. But people are desperate to belong even if it's hard—especially if it's hard.

In late 2023, the US surgeon general released a report titled "Our Epidemic of Loneliness and Isolation," which revealed half of American adults had been experiencing measurable levels of loneliness even before COVID lockdowns.[7] The report also highlighted the significant physical consequences of loneliness and isolation— a 29 percent increased risk of heart disease, a 32 percent increased risk of a stroke, and a 50 percent increased risk of dementia for older adults.

A biblical ecclesiology demands the church respond by inviting strangers to become members of a household, or family. Every lonely, isolated person in our midst is one open door away from this sort of belonging—the open door of our hospitality, kindness, empathy, vulnerability, compassion, and generosity. But this door is difficult to open digitally or walk through virtually. It's almost always an embodied experience.

It's far too easy online to craft and curate our experiences toward self-serving ends. Our screens demand nothing of us other than our undivided addiction. But they never ask us to commit or sacrifice or give ourselves for or to other embodied humans. Quite the opposite. They ask that we see and interact only with disembodied avatars who exist as supplemental background noise

---

7   "Our Epidemic of Loneliness and Isolation," US Department of Health and Human Services, 2023, https://www.hhs.gov/sites/default/files/surgeon-general-social-connection-advisory.pdf.

to the soundtrack of our autonomous, individual selves. Online engagement is staring into a mirror.

Postman wrote that "a mirror records only what you are wearing today. It is silent about yesterday," leading us "into a continuous, incoherent present" because experiencing reality in the present fully and richly "requires a contextual basis."[8] The mirror of television that Postman was writing about then has become the mirror of the smartphone, with one crucial difference. Where once the mirror was fixed in place, typically the living room or family room, now the mirror is in our pockets, with us everywhere we go.

The social psychologist Jean Twenge has tracked the shocking rise in teenage anxiety and depression, pinpointing 2012 as the year when we began to see rates skyrocket. That was the year smartphone ownership in America tipped over 50 percent and teens first began owning their own. It was also the year social media went from optional to culturally mandatory. Then as an aside, Twenge notes that 2012 was also the year when smartphones began featuring front-facing cameras.[9] Our phones became literal mirrors. The next year, unsurprisingly, *selfie* was Oxford Dictionary's word of the year. The proliferation of mobile mirrors led people to experience life more and more as a performance, with themselves as the star.

By contrast, analog presence in a local church invites us to look not at ourselves but at others, becoming an embodied context for one another. When "church" is reduced to staring at a screen, the mirror simply reflects our preferences and proclivities. But gathering

---

8   Postman, *Amusing Ourselves*, 137.

9   Derek Thompson, "Why Americans Suddenly Stopped Hanging Out," *Atlantic*, February 14, 2024, https://www.theatlantic.com/. See also Jean Twenge, i*Gen: Why Today's Super-Connected Kids Are Growing Up Less Rebellious, More Tolerant, Less Happy—and Completely Unprepared for Adulthood* (New York: Atria Books, 2017).

together shatters the mirror and helps us see beyond the abyss of self and out onto the expansive horizon of communal life in God's kingdom. And when we look there long enough, we discover its life-giving power. Gathering is good for us.

According to the Center for Disease Control, embodied social connectedness decreases risk of premature death, reduces anxiety and depression levels, and increases mental and emotional health. Those who show up regularly to gather with others display lower rates of cardiovascular problems and higher rates of cognitive function and immune health.[10] Singing together with others reduces stress hormones and increases cytokine proteins, which aid in the body's resilience against serious illness. Corporate singing also stimulates memory, increases lung capacity, and eases postpartum depression.[11] The God who breathed life into dirt and made embodied humans breathes life into us today through embodied community.

A year into COVID lockdown, our county had finally lifted restrictions enough that our church could gather in a large group outdoors. On Easter Sunday 2021, hundreds of us gathered to celebrate the resurrection under large tents in our parking lot. We sang, listened, and laughed together. We couldn't get people to leave. I didn't want to leave. Not much was different from what we'd been receiving online. The song quantity was the same. The preaching quality was the same. But everything was different. The digital walls had been removed, and we'd now walked through embodied doors. We were truly together. It was messy and strange and awkward and

10 "How Does Social Connectedness Affect Health?," Center for Disease Control and Prevention, March 30, 2023, https://www.cdc.gov/.

11 Alexandra Moe, "Singing Is Good for You. Singing with Others May Be Even Better," *Washington Post*, June 25, 2023, https://www.washingtonpost.com/.

inconvenient. It was beautiful and profound and comforting and healing. After months of scrolling ourselves to death, we found new life in the real presence of real people in a real place.

Day by Day

"When news is packaged as entertainment," Postman wrote, we lose "our sense of what it means to be well informed."[12] In much the same way, when "church" is packaged as "content," which it invariably is online, we lose our sense of what it means to be *transformed*. To be a Christian is to be on a lifelong journey of transformation into Christlikeness alongside a community of others pursuing the same. It's not so much that digital ministry and online church are bad; it's that they are inadequate and incomplete. Watching can inform us *about* and sometimes inspire us *toward* God's people. But only embodied participation can transform us *into* God's people.

When pastors and church leaders give inordinate energy toward online engagement, unintentionally conceding our ecclesiology to the comfort and convenience of digital platforms, we reshape congregations into audiences. Teaching gives way to entertainment. Communion gives way to commercialization.

The way forward, then, may be to look for guidance in the earliest days of the Jesus movement. Without romanticizing the early church, as many are prone today, what we see there is nothing less than an inconveniently high commitment to one another: "They *devoted* themselves to the apostles' *teaching* and the *fellowship*, to the *breaking of bread* and the prayers" (Acts 2:42).

At the start, the congregating people of God devoted themselves to teaching and collective communion, the breaking of bread. This

---

12 Postman, *Amusing Ourselves*, 107–8.

is what constituted the church. In the digital age, our devotions still run deep but in vastly different directions. Pastors are now tasked with the difficult but vital work of inviting those we serve to reorient their understanding about and rehabituate their desires for the local church.

Left unchecked, digital ministry and online church are forming us into audiences yearning to be entertained by consumer-friendly content. Our pastoral task is to invite people to once again be formed into congregations yearning for teaching and communion, with God and his people. This is when the truly spectacular is most possible: "Awe came upon every soul, and many wonders and signs were being done through the apostles. And all who believed were together and had all things in common. And they were selling their possessions and belongings and distributing the proceeds to all, as any had need" (Acts 2:43–45).

The first Christians experienced awe and encountered wonders and signs, but it all took place within the context of a people who "were together and had all things in common," sacrificing to provide for one another's needs. It was the miraculous in the middle of the mundane. The rest of Acts makes clear that the early church had its fair share of issues, tensions, and complexities. And yet, the movement took hold. Maybe one of the reasons why is that despite the challenges, they pushed through the inconvenience and gave themselves to the high-commitment calling of showing up, over and over again: "And *day by day*, attending the temple together and breaking bread in their homes, they received their food with glad and generous hearts, praising God and having favor with all the people. And the Lord added to their number day by day those who were being saved" (Acts 2:46–47).

Day by day. I wonder sometimes if these men and women kept showing up every day to be together because their implausible origin story as a community began that way—all of them gathered in one place: "When the day of Pentecost arrived, they were all together in one place. And suddenly there came from heaven a sound . . ." (Acts 2:1–2).

We know how the rest of the story goes. The church is born and history is changed. Proponents of digital ministry and online church are fond of arguing that if Jesus or Paul had had the internet, they would've been all in and maximized the technology to take the gospel far and wide. Maybe they're right. But I'm certain the Pentecost story could not and would not have transpired quite like this over Zoom.

Everyone is together in one place. God shows up. And an unlikely group of people are knit together as one. Amid linguistic diversity, the people hear the same gospel words. This is the reversal of Babel. Where once division reigned, Pentecost unites disparate individuals into a family.

This story echoes through all generations, even one as digital and divided as ours. This is why it matters so much, especially now, that we continue gathering together, offering one another the gift of our presence.

Discussion Questions

1. How would you answer Kim's key questions at the beginning of the chapter: "What is the problem that digital ministry and online church solve? What new problems do we create by solving this problem?"

2. Kim argues that "inconveniently high commitment" helped bring the early church together. Have you seen this sort of commitment in the churches you have been part of? How might contemporary

technologies condition people to resist robust commitment? What is it about "inconvenience" that might lead people toward Jesus Christ and not away from him?

3. Have you seen loneliness connected with screen time in your own life? How can screen usage exacerbate loneliness and, inversely, how can loneliness exacerbate screen usage?

4. Kim argues, "When 'church' is packaged as 'content,' which it invariably is online, we lose our sense of what it means to be *transformed*." After reading this chapter, how would you explain the spiritual significance and theological necessity of church as the real presence of real people in a real place?

14

Heed Huxley's Warning

Andrew Spencer

IN THE EARLY 2000S, MIT graduate student Eric Brende lived off grid in an Amish community as a year-long experiment. As newlyweds, Eric and his wife, Mary, moved into a home with no electricity, determined to evaluate the importance of modern technology in their lives. Although their experience was overwhelmingly positive, the couple abandoned the countryside. Unfortunately, Mary is allergic to horses.[1]

They now live in a city but intentionally limit their use of technology. Their posture is one of cautious resistance, not outright rejection, and they are convinced this approach has improved their quality of life. Their story is a curiosity, but it points to the potential for modern humans to evaluate technologies rather than simply and unconsciously adopting them.

---

1 Eric Brende, *Better Off: Flipping the Switch on Technology* (New York: HarperCollins, 2004).

Postman's major concern in *Amusing Ourselves to Death* was that people adopted television without considering its negative effects on how we think and communicate. Postman concluded with a "Huxleyan warning" that "in an age of advanced technology, spiritual devastation is more likely to come from an enemy with a smiling face than from one whose countenance exudes suspicion and hate."[2] George Orwell's militaristic nanny state from *1984* seems less likely to be forced on the free world than the voluntary acceptance of the nightmare of Aldous Huxley's drug-saturated *Brave New World*. According to Postman, Huxley's dystopia is more likely, less pleasant, and more difficult to escape.

Despite Postman's stark warning and prescient analysis, *Amusing Ourselves to Death* ended in a whimper of application. His main recommendation was simply to promote education in media ecology, teaching children "how to distance themselves from their forms of information" so they understand "the politics and epistemology of media."[3] He was right that the way forward required education to raise awareness, but education alone is insufficient to turn the tide of the soft tyranny of technology. We have to dig deeper.

Widespread resistance to technological domination can happen only when robust communities with strong social fabrics enable that resistance. Because they are culturally embedded but subject to the supreme authority of Scripture, Christian communities have the resources to resist the negative influences of technology and break the self-reinforcing feedback loops a tech-saturated society creates. We have the gospel's hope to guide us beyond the utopic technological visions of our day.

2 Neil Postman, *Amusing Ourselves to Death*, 20th anniversary ed. (1985; repr., New York: Penguin Books, 2005), 155.
3 Postman, *Amusing Ourselves*, 163.

The riddle of modern life is determining the way to resist technology's corrosive effects while still living in a tech-saturated world. There are no easy answers to this riddle, but we can learn—as Eric Brende did—from religious communities like the Amish who evaluate and adopt technologies based on each one's influence on their social fabric. Their example helps us understand the way technology exerts control over a culture, which in turn helps us see some practical steps Christians can take to avoid the dystopian future Postman warns about.

Living in an Amish Paradise?

When we hear the word *Amish*, most of us think of barn raisings, horses and buggies, or the sensationalized tradition of Rumspringa. These are curiosities that distract from what we can really learn from their community. Amish communities are one of the few demographic groups that has consistently doubled in size roughly every twenty years.[4] This growth is due to their high birth rate and, despite their lifestyle's cultural strangeness, an exceptionally high retention rate. They do not recruit members into their community or faith. Therefore, it's not their particular way of life, and certainly not their theology, that can be most helpful for evangelicals. Rather, it's that the community's strength is the primary concern when they evaluate technology.

People often associate "Amish" with being technologically primitive. As Donald Kraybill, a professor of sociology and Anabaptist studies, notes, "Some stereotypes of Amish life imply that they reject technology and live in a nineteenth-century cocoon. Such images are false. The Amish adopt technology

---

4   Sam Myers, "Across the Country, Amish Populations Are on the Rise," Daily Yonder, April 10, 2024, https://dailyyonder.com/.

selectively, hoping that the tools they use will build community rather than harm it."[5]

The Amish use technology rather than being used by it. Each Amish community navigates its technology adoption in a localized, community-oriented process. For example, some communities allow gas engines to power machinery on horse-drawn farming equipment. Some Amish merchants use cell phones. Meanwhile, they generally contract their non-Amish neighbors to drive them in vans when they need to get somewhere quickly. Outsiders sometimes marvel at the apparent inconsistency in Amish technological choices. But it's a reflection of the careful evaluation of technologies on a case-by-case basis.

The important fact is not *which* technologies an Amish community adopts but rather *that* they purposefully evaluate them. "Clearly, they want to build community and preserve social capital," Kraybill argues, "so they employ technology cautiously, ever wary of its potential to tatter the social fabric."[6] The process of evaluating technologies for adoption as a community has largely protected the Amish from the socially isolating effects of smartphones and other digital tech.[7] As we consider Postman's Huxleyan warning about technology and apply that warning to our churches and families, we can borrow aspects of their screening process for our own families and communities, though not without sifting the wheat from the chaff.

Technological decisions are made in a top-down fashion among the Amish in a manner that will not work well in most evangelical

5   Donald B. Kraybill, *The Riddle of Amish Culture* (Baltimore: Johns Hopkins University Press, 2001), 188.

6   Kraybill, *Riddle of Amish Culture*, 212.

7   Sherry Turkle explores these effects in *Alone Together: Why We Expect More from Technology and Less from Each Other* (New York: Basic, 2011).

churches. Community norms are enforced, under threat of ex-
clusion from the community, with significant intrusion into the
private lives of community members. The Amish bishop has the
authority within the community to order bathrooms removed
from homes, rubber tires taken off wagons, or electric wires ripped
out even after they have been installed at great expense.[8] That
level of authoritarian control is inconsistent with the priesthood
of all believers.

Despite the flaws of their approach, the Amish communities'
rich social fabric has a powerful attraction in our increasingly
fragmented world. To gain the benefits of Amish culture without
the flaws, we have to understand what they are effectively resisting,
which is the content of Postman's warning.

Accelerating Tyranny of Technopoly

Postman warned that culture will be deformed if people voluntarily
adopt technologies without considering the negative consequences,
which is how Huxley's *Brave New World* came to be. He argued
that we were susceptible to this because "public consciousness has
not yet assimilated the point that technology is ideology."[9] In other
words, technology is not just a product of culture; it also shapes
culture in unexpected ways.

Moreover, technology tends to reshape culture faster than
we expect. For example, Postman underestimated the transformative power of computers, which he called "vastly overrated
technology."[10] In 1985, the Macintosh computer had been available
only for a year. He could not have imagined everyone carrying an

8   Kraybill, *Riddle of Amish Culture*, 297–302.
9   Postman, *Amusing Ourselves*, 157.
10 Postman, *Amusing Ourselves*, 161.

internet-connected supercomputer in his or her pocket. Yet the acceleration of technological progress illustrates the ideological power it can have within a culture.

Less than a decade after *Amusing Ourselves*, Postman wrote *Technopoly: The Surrender of Culture to Technology*. The sequel was necessary because he saw his Huxleyan predictions rapidly becoming reality. He explained,

> Technopoly eliminates alternatives to itself in precisely the way Aldous Huxley outlined in *Brave New World*. It does not make them illegal. It does not make them immoral. It does not even make them unpopular. It makes them invisible and therefore irrelevant. And it does so by redefining what we mean by religion, by art, by family, by politics, by history, by truth, by privacy, by intelligence, so that our definitions fit its new requirements. Technopoly, in other words, is totalitarian technocracy.[11]

We have many examples of technopoly in our culture, but one of the most obvious is the smartphone. It's commonly maligned because of its apparent correlations with deteriorating mental health in younger generations.[12] Yet the cultural changes introduced by mobile telephony go much further.

Consider how even the basic mobile phone has changed our infrastructure. Pay phones have largely been removed from public spaces because most people don't need them. There are no laws prohibiting them, but when the majority have a phone in their

11 Neil Postman, *Technopoly: The Surrender of Culture to Technology* (New York: Knopf, 1992), 48.
12 Jonathan Haidt, *The Anxious Generation: How the Great Rewiring of Childhood Is Causing an Epidemic of Mental Illness* (New York: Penguin, 2024).

pockets, pay phones are unprofitable. But their gradual disappearance from public spaces reinforced the need for nearly everyone to acquire his or her own mobile phone. In the past, for example, the pay phone at school was a reliable means for a child to arrange a pickup after a school activity was complete. Now parents feel obligated to provide their children with phones for the purposes of convenience and safety. This is technopoly.

No cultural committee or legislature met to mandate mobile phone adoption. The change took place incrementally, voluntarily, and without much public debate. In just a few decades, the mobile phone has gone from being a luxury tool we can use to make life easier to a technology so widespread that many everyday functions can't happen without it. This is technopoly. It is dangerous because it creates a self-reinforcing feedback loop that requires an external input to disrupt it.

Breaking the Feedback Loop

Technopoly exists when a culture no longer simply uses the tools it develops but rather is being ideologically shaped by them, unawares.[13] We live in a technopoly. As with mobile phones, new technologies arise and are adopted with little consideration of their social costs. Some new technologies become necessary to address problems caused by existing technologies that came about through technopoly. Most significantly, many of our modern technologies unravel the social fabric necessary to resist technologies. Thus technopoly creates a self-reinforcing feedback loop. By isolating individuals from communities, technopoly neuters the threat of organized resistance.

---

13 Postman, *Technopoly*, 13.

We've all experienced the painful squeal of a sound system when a microphone is placed in front of a speaker. That's a self-reinforcing feedback loop. The sound tech has two options in the mad scramble to stop the squeal: mute the input or the output. Without the sound tech, the noise can become unbearable and the speakers may be damaged. Someone needs to notice the problem and do something about it.

To break the feedback loop of a technopoly, some people within the culture must be aware of the problem and be willing to take action to resist. Because of the strength of our community ties through the gospel, Christians are equipped to address the accelerating effects of technopoly in the broader culture if we take positive steps to break the feedback loop.

In *Technopoly*, Postman offered two proposals to prevent a Huxleyan future. At the individual level, he argued each person "must try to be a loving resistance fighter."[14] At the communal level, he proposed educating people to raise media awareness and to inoculate them against the negative effects of technology. Both ideas can be applied to local congregations tempted to scroll themselves to death.

Become Technological Resistance Fighters

Technological resistance fighters live so that technology "always appears somewhat strange, never inevitable, never natural."[15] One way to encourage this distance is to refuse to use certain technologies. In *Better Off*, Eric Brende offers an extreme example of resistance. More moderately, Andy Crouch recommends adopting a "minimal pattern of Sabbath: we choose to turn our devices off *not just one day every week, but also one hour (or more) every day and one week*

---

14 Postman, *Technopoly*, 182.
15 Postman, *Technopoly.*, 185.

(or more) every year."[16] Explicit and firm limits help believers resist technopoly.

Another approach comes from Brett McCracken's nutritional model in *The Wisdom Pyramid.*[17] He recommends focusing on using technology purposefully and thoughtfully, much like the way we eat a variety of foods to maintain a healthy diet. McCracken's paradigm has the advantage of focusing on positively replacing electronic media technologies with Scripture intake, church participation, time in nature, book reading, and enjoyment of beauty rather than simply trying to resist the dopamine hits of social media by brute force.

Individuals and families must become more careful consumers of technology and media content by maintaining their awareness of the foreignness of these technologies, but it'll be difficult to do so without community support. This is where the local congregation can step in.

A congregation that encourages using paper Bibles during services, discourages using phones during group gatherings, discusses the effects of innovative technologies, and holds one another accountable for social media use will be on the right track. Positive steps like these form loving resistance fighters. Yet we can't assume most Christians naturally see connections between their faith and the technologies they use in everyday life, nor that they even have a vocabulary for thinking Christianly about something like technology. The will to establish technological limits within a congregation's community norms must be fostered through education.

---

16 Andy Crouch, *The Tech-Wise Family: Everyday Steps for Putting Technology in Its Proper Place* (Grand Rapids, MI: Baker Academic, 2021), 98. Emphasis original.
17 Brett McCracken, *The Wisdom Pyramid: Feeding Your Soul in a Post-Truth World* (Wheaton, IL: Crossway, 2021).

Educate within the Church

Sunday school was invented in the eighteenth century to educate child factory-workers in basic academic subjects. We've since adapted the model to focus nearly entirely on Bible study, in large part because our public education system has been effective at creating a literate population. There are signs the system is fraying, as Postman warned in *Amusing Ourselves*.

Therefore, congregations should broaden their Sunday school curriculum from pure Bible instruction to helping believers make sense of the world. Biblical literacy has never been more important. But we must also recognize how the Bible interacts with all genres of human knowledge. In a post-Christian world, we may need to expand our curricular focus because, as Richard Lovelace argues, "Biblical truth is not a compendium of all necessary knowledge, but a touchstone for testing and verifying other kinds of truth and a structure for integrating them."[18]

The church must show Christians of all ages how to test and integrate truth into a Christian worldview, which requires expanding the scope of our discipleship programs to include topics like technology. This, in turn, will motivate believers to turn the technological tide with appropriate urgency.

Change the Brave New World

Writing in 1931, Huxley imagined his *Brave New World* dystopia in the distant future. The novel was set in the year 2450. However, by 1958 Huxley wrote, "The prophecies made in 1931 are coming true much sooner than I thought they would."[19] Mere decades

18 Richard Lovelace, *Dynamics of Spiritual Life: An Evangelical Theology of Renewal* (Downers Grove, IL: IVP Academic, 1979), 219.
19 Aldous Huxley, *Brave New World Revisited* (New York: HarperPerennial, 1958), 2.

later, I am amazed by the pace at which technology changes our culture. Evidence continues to accumulate about the negative effects of many technologies on culture. Though we need not be breathless or angry about the changes, becoming loving resistance fighters against technopoly shows that something beyond cultural conformity matters to Christians. The difference in lifestyle offers a window into our hope.

Eric and Mary Brende continue to live a low-tech, alternative lifestyle in a historic home in a historic section of St. Louis, within sight of the Gateway Arch.[20] It can be done. If evangelicals recognize technology's negative effects and actively resist the distortions it introduces into our lives, we will personally benefit and stand out from the world around us. If we keep technologies like smartphones at the periphery of our congregational life, we highlight the distinct humaneness of a Christian community in an inhuman cultural context. And that can be the first step in changing the world.

Seek the Good of the City

Christianity is a transcultural religion and can, to some degree, be contextualized within any culture. The challenge is that Christians must constantly evaluate their cultural contexts. The primary influence on our perspective—the authority of Scripture—means we have a source outside the culture itself to guide us as we evaluate that culture. Thus, Christian communities have the resources to break the feedback loop of technopoly.

The greatest danger to Christians is becoming blind to our culture's flaws. Such blindness can lead believers into invisible

20 The Brendes maintain a website that talks about some of their ongoing life: https://thehappy householder.com/.

patterns of sin. According to Lovelace, the most common form of temptation used by Satan is "involving believers in whole ways of life or patterns of behavior which are subchristian, which will extinguish their spirituality and make them negative witnesses; or luring them into adopting outlooks which excuse or justify sin and which may almost totally obscure their faith."[21] More and more, these subchristian "ways of life" and "patterns of behavior" are unconsciously formed by technology.

We need not become technologically backward to correct course. Yet we do need to become aware of how technology shapes culture and creates ways of living that undermine spiritual and ecclesial health. We must become loving technological resistance fighters as an outworking of our call to seek the good of the city in which we dwell (Jer. 29:7).

We should heed Postman's warning. The forty years since *Amusing Ourselves* have proved the prescience of his analysis. He was right to argue that we must become loving resistance fighters, but his worldview lacked the resources to explain how to make that happen. However, Christians have the resources within our historic faith to resist the technopoly Huxley depicted. But resisting what's unhealthy is just the starting point. We must also guide others toward *restoration* through habits of health.

After all, the church's mandate is not just to sound the alarm and stop our neighbors from scrolling themselves to death. We must also invite them into *life*—the abundant life offered in Christ (John 10:10). It's a life more abundant than all the conveniences of algorithms, more satisfying than all the pleasures of our smartphones, more beautiful than all the wonders we could possibly scroll across.

---

21 Lovelace, *Dynamics of Spiritual Life*, 137.

Discussion Questions

1. Consider the Amish. When evaluating whether or not to buy and use technology, is "the community's strength" one of your primary concerns? Do contemporary digital technologies market themselves for the strength of the community or the convenience of the individual?

2. Reflect on Andy Crouch's simple recommendation: choose to turn your devices off not just one day every week but also one hour (or more) every day and one week (or more) every year. Does this feel feasible to you? How does "technopoly" seek to prevent us from going without technology even for relatively short periods of time?

3. Postman calls our age a "totalitarian technocracy," which Spencer further explains as a society where alternate options to the dominant technology are becoming not just impractical but "irrelevant." How have you experienced this? Are there technologies you would like to go without in theory but have found it practically impossible to live without? If so, what are creative ways you could resist this oppressive sense of "not being able to live without" certain tech?

4. *Scrolling Ourselves to Death* gives us a lot to think about. What are three primary points, phrases, or practices you want to remember and/or implement in your daily life?

Epilogue

Ivan Mesa

ONE OF THE MANY REASONS I benefit from studying church history is that I'm disabused of the notion that we need to return to some pristine past. While I'm admittedly pessimistic about what the coming decades may bring, I'm no romantic who wants to turn back the clock to a "golden age" of yesteryear.[1] Even Postman fell prey to this temptation in his otherwise stellar work. His nostalgic enthusiasm for Enlightenment reason and the bygone "typographic age" sometimes limited his ability to speak prescriptively into the future—at least beyond "let's mimic eighteenth century reading and oratory habits."[2]

---

1 A good book to read on this topic is Alan Jacobs, *Breaking Bread with the Dead: A Reader's Guide to a More Tranquil Mind* (New York: Penguin, 2021). See also Robert Tracy McKenzie, *A Little Book for New Historians: Why and How to Study History* (Downers Grove, IL: IVP Academic, 2019); John Fea, *Why Study History?: Reflecting on the Importance of the Past*, 2nd ed. (Grand Rapids, MI: Baker Academic, 2024).

2 In 1999, President Bill Clinton made a call to build a bridge into the twenty-first century. Postman published a book the following year in which he called for building a bridge to the eighteenth century. See Neil Postman, *Building a Bridge to the 18th Century: How the Past Can Improve Our Future* (New York: Vintage, 2000).

Amazon founder Jeff Bezos said he's often asked to predict what will change in the next ten years.[3] In the hypercompetitive landscape of Silicon Valley, it's easy to see why that kind of foresight is perceived to be advantageous. Constant change and disruption have become so ingrained as cultural values—especially in the tech start-up world—that we assume "innovation" necessarily involves discontinuity more than continuity. To succeed, we think, we must blaze the trail to a radically new future before anyone else does.[4]

It's interesting, though, that Bezos responded and said he almost never gets the question "What's *not* going to change in the next ten years?" He then added, "And I submit to you that that second question is actually the more important of the two—because you can build a business strategy around the things that are stable in time." I suspect this level of contrarian focus is what has led to Amazon's dominance over the past three decades. Rather than searching for new hacks, Amazon has been able to build on what they absolutely know customers want: lower prices and faster delivery.

When you clarify what matters most, it focuses creative energy on the things that make the greatest difference over time. Christians, of course, know a thing or two about this. After all, the church over two millennia has held fast to the same book, preached the same Christ, and contended for the same faith that was once for all delivered to the saints (Jude 3).

---

3   Variations of this reflection have appeared in different contexts, but one comes from an interview Bezos did with *Harvard Business Review*. See Julia Kirby and Thomas A. Stewart, "The Institutional Yes," *Harvard Business Review*, October 2007, https://hbr.org/. The version cited here is taken from a live interview: "2012 re:Invent Day 2: Fireside Chat with Jeff Bezos & Werner Vogels," YouTube, https://www.youtube.com/watch?v=O4MtQGRIIuA. Quotation starts at 4:29.

4   For an analysis of this trend, see Brett McCracken, "Disrupting Ourselves to Death," The Gospel Coalition, January 18, 2023, https://www.thegospelcoalition.org/.

In a fast-moving world where discontinuity has more cultural cachet than continuity, Christians might be tempted to spend inordinate energy stressing about what is changing or will likely change, while the "what hasn't changed" aspects of our faith get forgotten or downplayed. To be sure, it's vital we attend to what's changing around us. That's been one of the primary goals of this book. But it's Christianity's unchanging aspects that will best propel us forward in health, come what may.

I can't tell you what new social media platforms will emerge in the coming decades to monetize our attention.[5] I can't tell you what industries AI will upend or even if it'll present an existential threat to humanity.[6] I can't tell you how technology will further erode our sense of self, diminish our communities, or corrode our public discourse.[7] I can't tell you all the many ways that, under God's sovereign hand, the church will steward technological developments for the sake of mission.[8] I have opinions about all these and more, but at the end of the day they're just probable guesses.

5   See Tim Wu, *The Attention Merchants: The Epic Scramble to Get Inside Our Heads* (New York: Knopf, 2016); Shoshana Zuboff, *The Age of Surveillance Capitalism: The Fight for a Human Future at the New Frontier of Power* (New York: Public Affairs, 2019); Jenny Odell, *How to Do Nothing: Resisting the Attention Economy* (Brooklyn, NY: Melville House, 2019). For a Christian reflection on this topic, see Tony Reinke, *Competing Spectacles: Treasuring Christ in the Media Age* (Wheaton, IL: Crossway, 2019).

6   See Jacob Shatzer, *Transhumanism and the Image of God: Today's Technology and the Future of Christian Discipleship* (Downers Grove, IL: IVP Academic, 2019); Henry A. Kissinger, Eric Schmidt, and Daniel Huttenlocher, *The Age of AI: And Our Human Future* (New York: Little, Brown and Company, 2021).

7   See Chris Bail, *Breaking the Social Media Prism: How to Make Our Platforms Less Polarizing* (Princeton, NJ: Princeton University Press, 2021); Jason Hannan, *Trolling Ourselves to Death: Democracy in the Age of Social Media*, Oxford Studies in Digital Studies (New York: Oxford University Press, 2023).

8   For the best overview of this topic, see Tony Reinke, *God, Technology, and the Christian Life* (Wheaton, IL: Crossway, 2022).

What I can tell you, however, is what *won't* change in the next ten years—and, barring Christ's return, the next ten after that and the ten after that.[9] Sinners will still need the gospel's hope, and saints will still need to walk in a manner worthy of that gospel. The church, the bride of Christ, will still have the same divine mandate to "go . . . and make disciples of all nations, baptizing them in the name of the Father and of the Son and of the Holy Spirit, teaching them to observe all that [Jesus] . . . commanded" (Matt. 28:19–20). What won't change is that the church will work toward this end with the promise of the risen Lamb to be "with [us] always, to the end of the age" (28:20). Despite the cultural and technological headwinds we'll doubtless face in the future, we have assurance that even amid the uncertainty—technological or otherwise—the Lord will guide us.

If someone stumbles on this book forty years from now, it's likely much in these pages will at points—similarly to Postman's classic—appear quaint and outdated. Our goal has been to assemble an array of insightful thinkers to reflect on the threats and opportunities facing the church *today*. In the coming years, I suspect we'll increasingly need more resources like this to help the church faithfully navigate a new batch of technological changes. We'll continue to need clear, countercultural thinkers and brave prophets who, like Postman, sound necessary alarms and provide timely guidance. We'll need frequent reminders of how the ancient gospel can speak truth and shape life in our brave new world.

---

9   For a book that examines this same concept but from a non-Christian perspective, see Morgan Housel, *Same as Ever: A Guide to What Never Changes* (New York: Portfolio, 2023). Housel dedicates his book to the "reasonable optimists." He argues that for progress to happen, one needs a combination of pessimism and optimism, both aware of the threats and clear-eyed about the opportunities. Our goal in this book has been to acknowledge the unique challenges we face but also suffuse our project with a confident expectation that, by God's grace, the church will prevail.

Whether we're amusing ourselves to death, scrolling ourselves to death, or harming ourselves in some other future way we can't presently imagine, the source of true life will remain the same. As Simon Peter answered Jesus, "Lord, to whom shall we go? You have the words of eternal life" (John 6:68). Our Christian task was, is, and will remain—until the Lord's return—to point the dying to the Lord and giver of life.

Acknowledgments

THE EDITORS (Brett and Ivan) wish to thank Amy Kruis, Samuel James, Todd Augustine, Tara Davis, and the entire Crossway team for their collaboration on this project. They are partners in ministry who share our heartbeat for serving the church of the Lord Jesus.

We're also indebted to Cassie Watson and J. T. Reeves for their editorial assistance on the manuscript, as well as our colleague and chapter contributor Andrew ("Spence") Spencer for first sparking the idea of doing something to commemorate the fortieth anniversary of Neil Postman's *Amusing Ourselves to Death*. We're grateful to all the contributors for accepting our invitation to bring their wisdom to this book, for the benefit of the church and the world.

Special thanks to Collin Hansen not only for his enthusiastic support of this project and contribution of a chapter, but also for his persistent encouragement and mentorship to us personally over the years. Amid our work at the Gospel Coalition—which involves frequent encounters with the good, bad, and ugly of the internet age—he faithfully leads us to do our work with continual innovation even as we seek to be anchored to the gospel and rooted in local churches. This book is a culmination of many conversations we have almost every week at the Gospel Coalition.

Finally, we'd be remiss not to express special gratitude to the late Neil Postman (1931–2003), whose prophetic voice spoke clearly forty years ago and reverberates—with urgent relevance—still today. It's a privilege for us to amplify his voice for such a time as this.

Contributors

Joe Carter is a senior writer for the Gospel Coalition, author of *The Life and Faith Field Guide for Parents*, the editor of the *NIV Lifehacks Bible*, and coauthor of *How to Argue Like Jesus*. He also serves as an associate pastor at McLean Bible Church in Arlington, Virginia.

Nathan A. Finn is professor of faith and culture and directs the Institute for Transformational Leadership at North Greenville University. He also serves as Teaching Pastor at First Baptist Church of Taylors, South Carolina. He has authored or edited over a dozen books, including *A Handbook of Theology*, *Historical Theology for the Church*, and *History: A Student's Guide*. Nathan and his wife, Leah, live with their four children in Tigerville, South Carolina.

Collin Hansen serves as vice president for content and editor in chief of the Gospel Coalition, as well as executive director of the Keller Center for Cultural Apologetics. He hosts the *Gospelbound* podcast and has written and contributed to many books, most recently *Timothy Keller: His Spiritual and Intellectual Formation* and *Rediscover Church*.

Samuel D. James serves as developmental and acquisitions editor at Crossway. He is the author of *Digital Liturgies*. He and his wife, Emily, live in Louisville, Kentucky, with their three children.

Jay Y. Kim serves as lead pastor at WestGate Church in the Silicon Valley of California. He's the author of *Analog Christian, Analog Church*, and *Listen Listen Speak*. Jay is also host of the *Digital Examen* podcast and cohost of the *Making Space* podcast. He is a graduate of Fuller Theological Seminary.

Hans Madueme is professor of theological studies at Covenant College in Lookout Mountain, Georgia, and is on the editorial board for *Themelios*. He's the author and editor of several books, including *Defending Sin: A Response to the Challenges of Evolution and the Natural Sciences*. Hans and his wife, Shelley, live with their two children in northwest Georgia.

Brett McCracken is a senior editor and director of communications at the Gospel Coalition. He is the author of *The Wisdom Pyramid*. Brett lives in Santa Ana, California with his wife, Kira, and their three children. They belong to Southlands Church.

Ivan Mesa is editorial director for the Gospel Coalition. He's editor of *Before You Lose Your Faith* and coeditor of *Faithful Exiles*. He and his wife, Sarah, have four children, and they live in eastern Georgia.

Jen Pollock Michel is the author of *In Good Time, A Habit Called Faith, Surprised by Paradox, Keeping Place*, and *Teach Us to Want*. She writes and speaks regularly. Jen lives in Cincinnati with her family where they attend New City Presbyterian.

Patrick Miller is a pastor at The Crossing in Columbia, Missouri. He offers cultural commentary on the podcast *Truth over Tribe* and daily devotions on *Ten Minute Bible Talks*. He writes about Christianity and technology on his newsletter *Endeavor* and is the coauthor of *Joyful Outsiders* and *Truth over Tribe*.

G. Shane Morris is a senior writer at the Colson Center and host of the *Upstream* podcast. Shane is currently studying for his MA from Trinity Evangelical Divinity School. He and his wife, Gabriela, live with their four children in Lakeland, Florida.

Keith Plummer is dean of the School of Divinity and professor of theology at Cairn University in Langhorne, Pennsylvania. He is a fellow of the Keller Center for Cultural Apologetics and contributor to *Before You Lose Your Faith* and *The Digital Public Square*. He also hosts the *defragmenting* podcast.

Read Mercer Schuchardt is associate professor of communication at Wheaton College, author of *Media, Journalism, and Communication: A Student's Guide* and coauthor of *Understanding Jacques Ellul*. Schuchardt earned his PhD in media ecology at New York University in 2005, where he studied under Neil Postman. He and his wife, Rachel, live in Wheaton, Illinois, and have ten children.

Andrew Spencer is associate editor for books at the Gospel Coalition. He is the author of *Hope for God's Creation* and editor of *The Christian Mind of C. S. Lewis*. He is an elder at CrossPointe Church and lives in Monroe, Michigan, with his wife and three children.

Thaddeus Williams serves as associate professor of theology at Biola University. He is the author of *Confronting Injustice without Compromising Truth*, *God Reforms Hearts*, *Don't Follow Your Heart*, and *Revering God*. He resides in Orange County, California, with his wife and four kids.

General Index

Scripture Index

TGC | THE GOSPEL COALITION

The Gospel Coalition (TGC) supports the church in making disciples of all nations, by providing gospel-centered resources that are trusted and timely, winsome and wise.

Guided by a Council of more than 40 pastors in the Reformed tradition, TGC seeks to advance gospel-centered ministry for the next generation by producing content (including articles, podcasts, videos, courses, and books) and convening leaders (including conferences, virtual events, training, and regional chapters).

In all of this we want to help Christians around the world better grasp the gospel of Jesus Christ and apply it to all of life in the 21st century. We want to offer biblical truth in an era of great confusion. We want to offer gospel-centered hope for the searching.

Join us by visiting TGC.org so you can be equipped to love God with all your heart, soul, mind, and strength, and to love your neighbor as yourself.

TGC.org